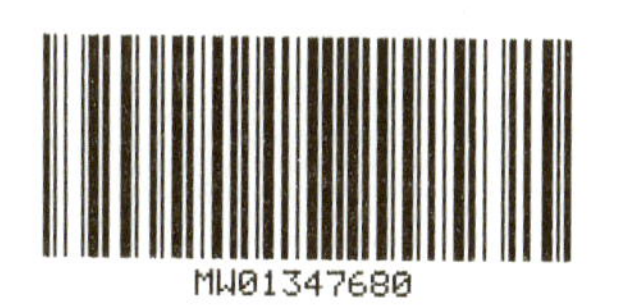

HEAT PUMPS

Operation • Installation • Service

Randy F. Petit, Sr. **Turner L. Collins**

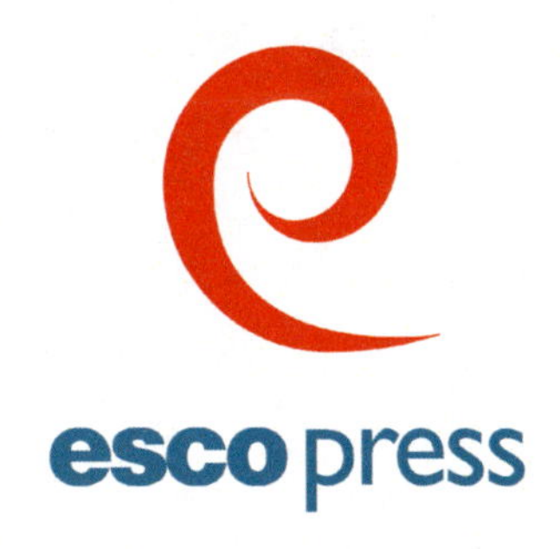

Mount Prospect, Illinois

ISBN 1-930044-29-1

Printed in the United States of America
7 6 5 4 3 2 1

Second Print April 2012

Table of Contents

Section 1: Basic Principles of Operation

Objectives

Upon completion of this section, the participant will be able to:

1. understand the history of heat pumps;
2. explain basic refrigeration terms and concepts;
3. identify the characteristics of different heat pump systems;
4. compare the three heat pump cycles/modes (cooling, heating and defrost).

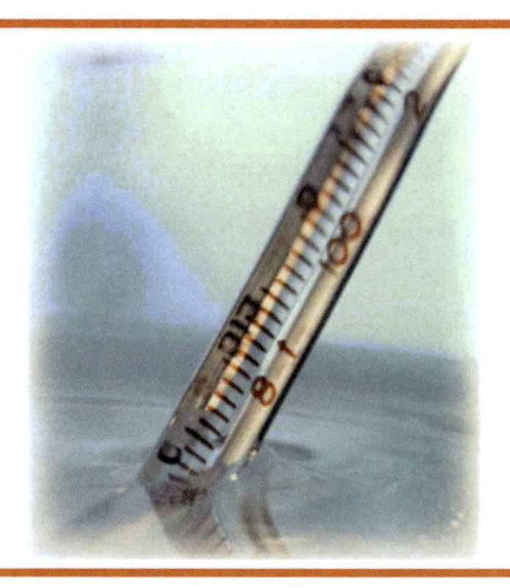

A Brief Heat Pump History

When heat pumps were introduced in the 1950s, very little training was provided for their installation and service. The first generation of heat pumps also had some design problems that increased their failure rate as compared to air cooling only systems. Customers were not satisfied with the benefits of heat pumps versus their high cost and poor operation, so heat pumps failed to become a vital part of the market.

Heat pumps were reintroduced during the energy shortage of the early 1970s. Manufacturers realized many improvements would have to be made before heat pumps could compete with conventional heating and cooling systems. Training became a top priority for air conditioning companies. Through training, technicians and installers gained a working knowledge of how to calculate heating and cooling loads, properly install air ducting and system components, and, just as importantly, maintain and service these systems.

Many remarkable improvements have been made since that time. Seasonal Energy Efficiency Ratio (SEER) ratings have been raised from around 7 to 14 and higher. Modern compressors are now much more dependable and efficient, with longer warranties available. Fan and blower motors provide variable speed control. With these vast improvements, service and maintenance problems have been greatly reduced.

Due to the adverse effects of the chlorine contained in many common refrigerants, heat pump manufacturers are now using more environmentally friendly refrigerants, such as hydrofluorocarbons (HFCs). Today's technicians must learn the characteristics of these refrigerants and the correct procedures for their use. The industry is constantly evolving and exploring new ways to develop more efficient and economical heat pumps. Modern heat pumps are reliable, functional and can truly be considered an asset to any home or business.

Basic Refrigeration Overview

Temperature

Temperature is most commonly thought of as the measurement of heat, however, temperature is actually the speed of the motion of molecules in a substance. The lowest possible temperature, -460°F, is called absolute zero. At absolute zero, all molecular motion ceases and all heat has been removed.

Temperature tells us how hot something is but not how much heat it contains. For example, if two containers, one containing one gallon of water and the other containing ten gallons of water, are placed over identical heat sources, the temperature in the one-gallon container will rise faster than the temperature in the ten-gallon container. Ten times the heat energy must be added to the larger container in order to heat the water in both containers to an equal temperature.

Measuring Temperature

Temperature is measured with a thermometer. Thermometers may be graduated in one of four different scales: Fahrenheit and Rankine are used in the United States, Celsius (Centigrade) and Kelvin are used in the metric system. On the Fahrenheit scale water freezes at 32°F and boils at 212°F. On the Celsius scale water freezes at 0°C and boils at 100°C.

For scientific calculations, absolute temperature scales without negative numbers must be used. On the Rankine scale, absolute zero is 0°R. The freezing point of water is calculated as 492°R (32°F + 460) and the boiling point of water is calculated as 672°R (212°F + 460).

On the Kelvin scale, absolute zero is 0°K. The freezing point of water is 273°K and the boiling point of water is calculated as 373°K(100°C + 273). The Kelvin scale is also called Celsius Absolute (CA).

Section 1: Basic Principles of Operation

Ambient Temperature

Ambient temperature is the temperature of the air surrounding an object. The components of a residential split air conditioning system are located both inside and outside the home. The temperature of inside components can be controlled, but ambient temperature outside the home can vary greatly, significantly influencing the efficiency of the system.

Dry-Bulb and Wet-Bulb Temperatures

Dry-bulb temperature refers to the actual air temperature and is measured with a dry-bulb (ordinary) thermometer. Wet-bulb temperature takes into account the amount of moisture in the air and is measured with a special thermometer that has a wick wetted with distilled water. Evaporation of the distilled water causes the wet-bulb thermometer to give a cooler reading than a dry-bulb thermometer. Less moisture in the air means that more evaporation occurs from the wick and therefore more cooling occurs; this results in a lower temperature reading. The difference between the dry-bulb and wet-bulb temperatures of the ambient air is called the wet-bulb depression. As the air's moisture content increases, there is less evaporation from the wick and less wet-bulb depression. When there is no wet-bulb depression, the relative humidity is 100 percent.

Dew Point Temperature

The temperature at which condensation forms is called the dew point temperature. Air can be dehumidified by passing it over a surface with a temperature below the dew point, such as an evaporator coil. Moisture collects on the cold surface of the evaporator coil and can then be drained through the condensate line.

Pressure

Pressure is defined as force per unit area and in the United States is most commonly described as pounds per square inch. The SI metric system uses a pressure unit called the pascal, named for French scientist Blaise Pascal. A pascal is equal to one newton per square meter (N/m2). A newton is the force required to cause a mass of one kilogram to accelerate at a rate of one meter per second.

The operation of an air conditioning or refrigeration system depends mainly upon pressure differences throughout the system. The pressure surrounding a substance affects its physical properties, especially its boiling point. Air conditioning technicians must be able to deal with pressures both above and below atmospheric pressure.

Atmospheric Pressure

Earth's atmosphere extends about fifty miles above the planet's surface and consists of approximately 78% nitrogen and 21% oxygen. The remaining 1% is composed of other

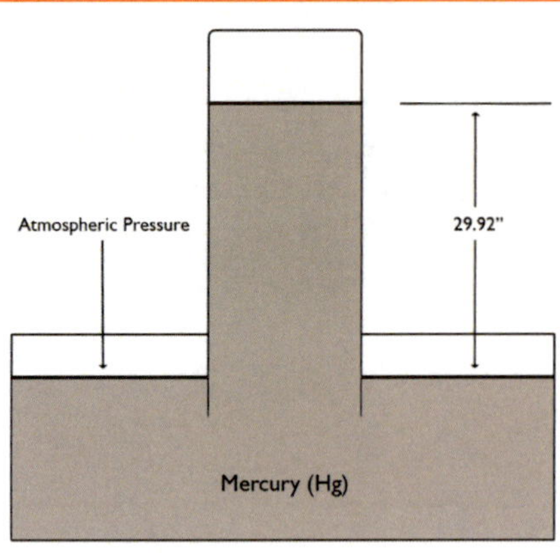

Mercury Barometer

gases. Even though gas molecules are very small, they still have weight. The atmosphere exerts a pressure of 14.7 pounds per square inch at sea level. At higher altitudes, atmospheric pressure is significantly less.

The most common way of measuring atmospheric pressure is with a mercury barometer. Normal atmospheric pressure at sea level (14.7 psia) will support a column of mercury 29.92 inches high.

Gauge Pressure

The pressure reading we most often use is called gauge pressure. Atmospheric pressure is shown as 0 psi or psig (pounds per square inch gauge).

Compound gauges used to measure low-side pressures in air conditioning systems can measure pressures both above and below 0 psig. Gauge readings are relative to atmospheric pressure. Compound gauge readings below atmospheric pressure are indicated in inches of mercury (Hg). It is necessary to adjust compound gauges periodically to compensate for changes in atmospheric pressure.

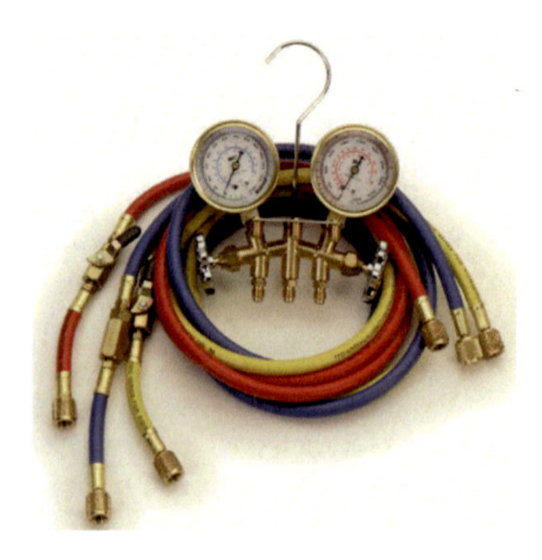

Vacuum

A thorough understanding of vacuum principles is a necessity for the technician. Pressures below atmospheric are usually read in inches of mercury (in. Hg) or millimeters of mercury (mm Hg). An increase in pressure increases the boiling point of a liquid and the opposite is also true. A lower pressure results in a lower boiling point. Any pressure below atmospheric is considered a partial vacuum. A perfect vacuum is the absence of all atmospheric pressure. A micron gauge is used for measuring deep vacuums. A micron is 1/1000 of a millimeter and one inch of mercury vacuum is equal to 25,400 microns.

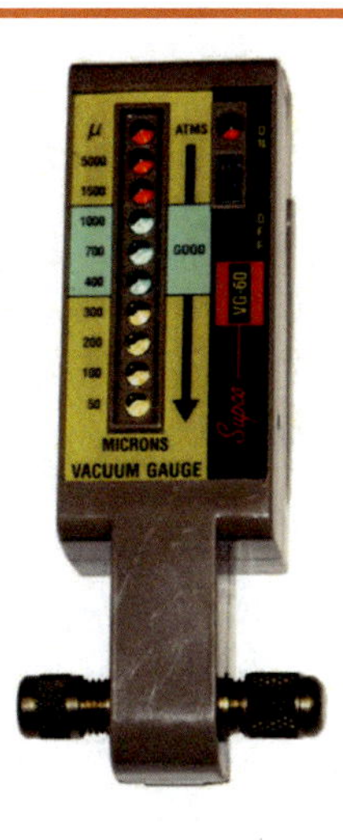

Absolute Pressure

The absolute pressure scale allows the measurement of both vacuum and pressure to be made using the same units. Absolute pressure measurements are indicated as psia (pounds per square inch absolute). 0 psia is the lowest possible absolute pressure.

Because atmospheric pressure is 14.7 psia at sea level, gauge pressure can be converted to absolute pressure by adding 14.7 to the gauge pressure measurement.

Thermodynamics

Heat is a form of energy. Refrigeration is the movement of heat from an area where it is not wanted to an area where it is less objectionable. For example, a refrigerator transfers heat from the inside of the cabinet to the outside. A heat pump is used to move heat in two directions.

The First Law of Thermodynamics

The laws of thermodynamics give us a basic understanding of heat. The first law of thermodynamics states that energy cannot be created or destroyed; it can only be converted from one form to another. This law applies to heat and to all forms of energy: electrical, mechanical, solar, chemical, and nuclear.

The Second Law of Thermodynamics

The second law of thermodynamics states that heat always travels from hot to cold. The speed at which heat travels from one object to another depends upon the temperature difference between the two objects. A greater temperature difference causes a faster heat transfer. The transfer of heat will stop when both objects reach the same temperature.

Since refrigeration involves the exchange of heat, it is important to understand how heat travels. There are three ways in which heat travels from one object to another: radiation, conduction, and convection.

Radiant heat from the sun warms objects.

Radiation

Heat is radiated by waves such as light waves and radio waves. Radiant heat waves are absorbed by solid objects, not by the space they travel through. For example, heat from the sun travels through space and then is absorbed by the earth and its atmosphere.

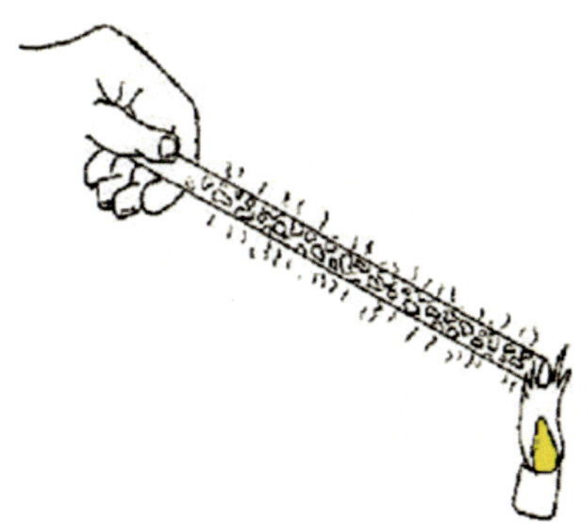

Heat is conducted directly through a metal bar with very little energy lost.

Conduction

Conduction is the transfer of heat either through an object or by direct contact between two objects. When one end of a metal bar is heated the heat travels by conduction to the cooler end of the bar.

Convection

Convection is the transfer of heat through fluids in either a liquid or vapor state. Convection creates currents as the warmer (expanding) fluid rises above the colder (contracting) fluid. Convection currents can be natural or forced. Natural convection occurs as warm air naturally rises above cooler air. Forced convection, when a fan or blower is used, allows the use of a smaller heat exchanger.

States of Matter

Matter can exist as a solid, liquid or gas, depending on pressure, temperature, and heat content. In all substances

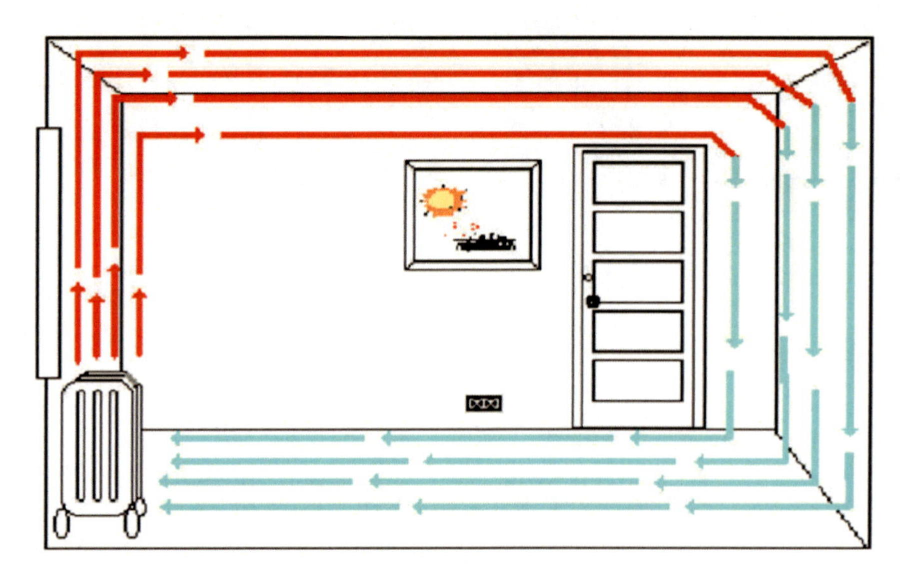

Convection current caused by warm air rising from the radiator is replaced by cooler air returning from the room.

above absolute zero molecules are constantly in motion. Even when matter is in its solid state, its molecules are moving. In the liquid state, molecules move faster. If liquid matter is heated further, its molecules move so fast that they are now free to travel in all directions as a gas.

Change of State

As a substance is heated or cooled, it may undergo a change of state. The substance most commonly used to exemplify this is water. At temperatures below 32°F, water exists in its solid form, ice. Between 32°F and 212°F water is a liquid, and at 212°F it becomes steam (at atmospheric pressure). The term used to describe any substance at its boiling point at a given pressure is saturation point. Material in saturated condition contains as much heat as it can absorb without changing state.

Atmospheric pressure has a direct influence on the saturation point (boiling point) of liquids. For example, water at sea level boils at 212°F; however, on a 15,000 foot high mountaintop, it boils at 184°F. Reduced atmospheric pressure at higher altitudes allows water to boil at a lower temperature.

Saturation point can be raised by increasing pressure. For example, placing a radiator cap on your automobile raises the saturation point of the water approximately 2.5°F for each pound of pressure increase.

When a material is undergoing a change of state, its temperature remains constant. In other words, when water boils, its temperature remains at 212°F until all of the water becomes steam. Adding heat will only make the water boil faster. During this process it is possible to have only liquid, only steam, or a mixture of liquid and steam.

This principle also applies when matter changes from gas to liquid (condensing) or liquid to solid (freezing). At 32°F it is possible to have only ice, only liquid, or a mixture of ice and liquid.

Since different materials have different molecular structures, the temperature at which a change of state occurs can vary. For example, the saturation point of water is 212°F at atmospheric pressure, but the saturation point of refrigerant HFC 410A is -57°F at atmospheric pressure.

Sensible Heat

Sensible heat is heat that is added to or removed from a material causing a change in temperature but not in state. Sensible heat can be felt with your hand or measured with a thermometer.

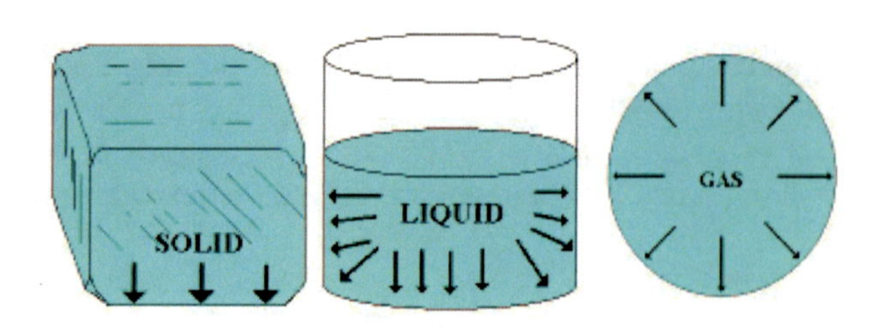

The three states of matter: solid, liquid and gas.

A thermometer measures only the temperature of a substance. It does not measure the amount of heat required to reach a certain temperature. Because a standard of measurement for quantity of heat was necessary, the international standard British Thermal Unit (BTU) was created. The BTU is a measurement of the exact amount of heat necessary to raise the temperature of one pound of water one degree Fahrenheit. If more than one pound of water is involved in a calculation, one BTU per pound of water is used. If more than one degree of temperature change is involved, one BTU for each degree of temperature change is used. This quantity of heat is called specific heat.

All materials are rated for specific heat using 1 for water as the base. The specific heat of air at standard conditions is 0.24 BTUs per pound.

Latent Heat

Latent heat is heat added to or removed from a substance causing a change of state but not in temperature. For example, water heated to 212°F may be either liquid or gas. To change liquid to gas, more heat must be added.

Remember, the temperature of water remains at 212°F as it boils. Additional heat is needed to change the water molecules from liquid to steam. Latent heat is sometimes called hidden heat because it cannot be felt by hand or measured with a thermometer.

There are five types of latent heat:
latent heat of melting
latent heat of fusion
latent heat of vaporization
latent heat of condensation
latent heat of sublimation

Every substance has a particular temperature and/or pressure point at which it changes state from solid to liquid or from liquid to vapor. At 212°F, 970 BTUs are required to change one pound of water into steam. To change steam into liquid, 970 BTUs must be removed.

This same principle applies to latent heat of fusion. At 32°F, 144 BTUs must be removed in order to freeze one pound of water. It takes exactly the same amount of energy (144 BTU per pound) to melt one pound of ice.

Latent heat of vaporization and condensation are the basic principles behind almost all refrigeration systems. As refrigerants absorb heat, they change from liquid to gas in the evaporator. The heat is then removed from the gas, changing it back into liquid in the condenser.

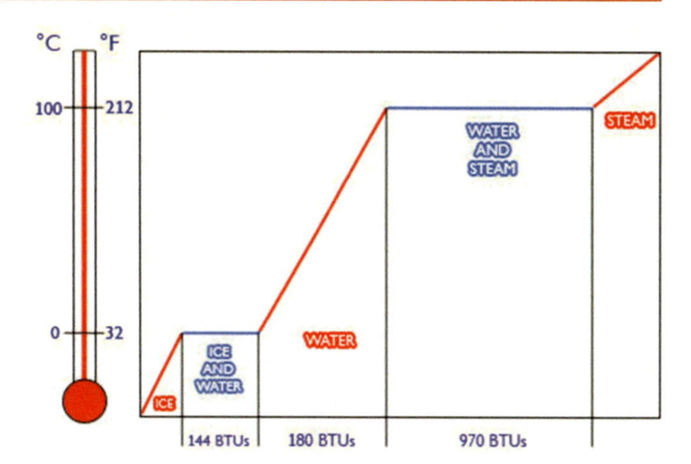

Changes of state for water: from solid to vapor

Some substances change directly from solid to gas with no visible evidence of changing into a liquid state. Dry ice (composed of carbon dioxide, or CO_2) is one such material. Latent heat of sublimation is responsible for this phenomenon.

Superheated Vapor

Superheated vapor is a gas that has been heated to a temperature above its saturation point (boiling point). When a substance is superheated, this means that sensible heat has been added after the substance has changed state. For example, if steam is heated to 213°F at atmospheric pressure, it is superheated 1°F.

The air around us consists mostly of nitrogen and oxygen. Both of these gases have boiling points well below 0°F at atmospheric pressure. Therefore, we can say that most of our atmosphere is made up of superheated vapors.

Subcooling

Any substance whose temperature is below its saturation point is considered subcooled. Water that is 210°F at atmospheric pressure is subcooled 2°F. In other words, subcooling is the number of degrees below the boiling point of a given substance.

Gas Law

A technician needs a thorough understanding of the behavior of gases in order to accurately read pressure gauges and diagnose system problems. Air conditioning systems constantly change liquid to gas and gas to liquid. The laws of gases deal with three factors: temperature, pressure and volume. Gas laws demonstrate that a gas will always behave according to the rules, with no exceptions.

Basic gas law formula: $P1 \times V1/T1 = P2 \times V2/T2$

Saturation Tables (Temperature/Pressure Charts)

A saturation chart is an extremely important tool for service technicians. Saturation tables illustrate the temperature/pressure relationship for refrigerants. Whenever pure refrigerant is contained in a system, a chart gives us the saturation temperature and pressure.

Refrigeration and air conditioning systems are designed to control a circulating refrigerant. In a basic refrigeration cycle, a refrigerant absorbs unwanted heat in one location and moves it to a place where it can be removed. The refrigerant absorbs heat through evaporation, then releases it through condensation. Vapor/compression systems are simple, but require a thorough understanding of theory and the operation of each component. A malfunction in one component can effect the operation of others in the system.

During a basic refrigeration cycle, refrigerant undergoes the processes of expansion, vaporization, compression, and condensation. Liquid refrigerant under high pressure is fed to a metering device. The metering device causes a drop in pressure as the refrigerant flows to the evaporator. This reduced pressure lowers the temperature of the refrigerant and causes it to vaporize as it absorbs heat.

An open cycle refrigeration system uses an expendable refrigerant (such as nitrogen), which is discarded after it has evaporated. Expendable refrigerants are found primarily in transport refrigeration systems. To reuse refrigerant, refrigerant vapor is drawn through a suction line and into the compressor. The compressed refrigerant is then pumped to the condenser, where it begins to cool. When the refrigerant is cool enough, it condenses and becomes liquid. It then travels through the liquid line to the metering device, and the refrigeration cycle is repeated.

High temperature refrigeration maintains a space temperature of between 47°F and 80°F. This is most often used to maintain human comfort or to preserve perishable food items or organic materials.

Medium temperature refrigeration maintains a box temperature of between 28°F and 40°F. The fresh food storage compartment of a domestic refrigerator is an example of a medium temperature application. Medium temperature is above freezing for most products.

Low temperature refrigeration maintains a temperature below 32°F. Most applications start at 0° F, and may maintain temperatures as low as -20°F.

Ultra low temperature refrigeration is also referred to as cryogenic. Cryogenic temperatures are between -250°F and absolute zero, or -459.7°F. Gases such as oxygen, nitrogen and helium can be liquefied at these low temperatures. A cascade design, which uses the evaporator of one system to cool the condenser of another system, can be used to achieve these low temperatures. Many stages may be required in a cascade system to reach these ultra low temperatures.

Temperature	Refrigerant Pressure		
	R-134a	R-22	R-410A
30°F	26.1	54.9	97.5
35°F	30.4	61.5	107.9
40°F	35.0	68.6	118.9
45°F	40.0	76.1	130.7
50°F	45.4	84.1	143.3
55°F	51.2	92.6	156.6
60°F	57.4	101.6	170.7
65°F	64.0	111.3	185.7
70°F	71.1	121.5	201.5
75°F	78.6	132.2	218.2
80°F	86.7	143.7	235.9
85°F	95.2	155.7	254.6
90°F	104.3	168.4	274.3
95°F	113.9	181.9	295.0
100°F	124.1	196.0	316.9
105°F	134.9	210.8	339.9
110°F	146.3	226.4	364.1
115°F	158.4	242.8	389.6
120°F	171.1	260.0	416.4

Temperature/Pressure Chart

System Components

Evaporator

An evaporator is a heat exchange device located in the area where cooling is desired. Liquid refrigerant and some flash gas is fed into the evaporator at a reduced pressure through the metering device. Once in the evaporator, liquid refrigerant changes state by absorbing the latent heat in the air passing over the coil. The lower temperature of the evaporator causes the moisture in the air to condense.

A drain captures the condensation and a trap in the drain line prevents air and debris from being pulled into the unit. Proper airflow and clean coils must be maintained in order to insure adequate heat exchange and efficiency.

Air conditioning systems are designed so that all refrigerant vaporizes in the evaporator. Only superheated gas leaves the evaporator. Airflow should be approximately 400 cubic feet per minute (cfm) for an air conditioning evaporator to operate correctly. Dirty filters or improper fan operation can reduce the amount of heat absorbed by the refrigerant in the evaporator and may cause flooding of liquid refrigerant back to the compressor. Since liquids are essentially incompressible, liquid refrigerant allowed to reach the compressor causes serious damage. During the heating cycle, the indoor coil is used as the condenser, which requires a higher volume of air. Air flow may be increased up to 450 cfm for a heat pump indoor air handler.

The amount of heat that an evaporator can absorb is equal to the amount of heat radiated by the condenser, minus the heat of compression. Multiple evaporators are sometimes used to provide zoned areas of cooling or for multi-staged cooling and heating.

Compressor

There are several different compressor designs but each serves the same purpose. By creating a pressure difference in the system, a compressor performs two functions. The first is providing suction to maintain low pressure in the evaporator. The second is providing compression to convert low pressure/low temperature refrigerant in the evaporator to high pressure/high temperature vapor, or highly superheated vapor. This highly superheated gas is then sent to the condenser, where another heat exchange process takes place.

The three most common compressor types found in residential and light commercial air conditioning are reciprocating, rotary and scroll.

A reciprocating compressor compresses refrigerant using pistons on a crankshaft. The reciprocating compressor is a positive displacement device that creates a vacuum in the cylinder, drawing refrigerant vapor into the cylinder on the down stroke and then compresses it on the upstroke, discharging it through a discharge port. There is a small clearance space at the top of the cylinder, which always holds a small amount of refrigerant. (A large clearance space would lower the compressor's capacity.) When a vacuum test is performed, a reciprocating compressor should be capable of pulling 26 to 28 inches of mercury (in. Hg) when discharging to the atmosphere or 24 inches of mercury when pumping against 100 psig-discharge pressure.

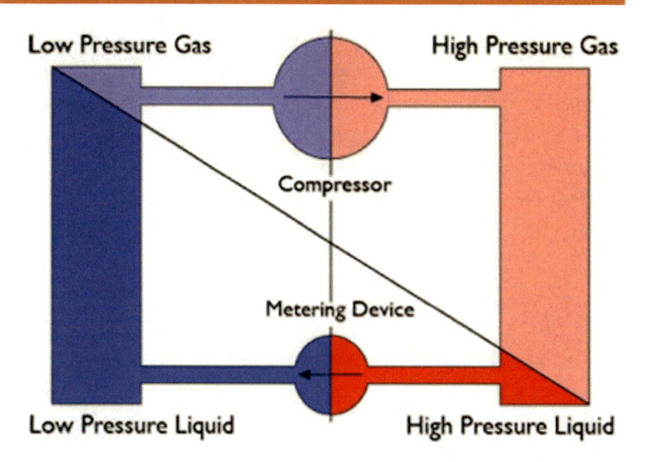

Basic Refrigeration Cycle

Because liquids cannot be compressed, catastrophic failure can result if liquid refrigerant ever enters the compressor.

A rotary compressor has fewer moving parts than a reciprocating compressor. In a rotary compressor, a drum-like piston rotates inside a chamber, squeezing refrigerant out through a discharge port.

A scroll compressor utilizes a mechanism resembling two flat coiled springs—one stationary and one connected to the motor shaft. The movable coil orbits inside the stationary coil and squeezes low-pressure refrigerant toward a high-pressure outlet. Large commercial systems use either screw or centrifugal compressors capable of moving large amounts of refrigerant through a system. Screw compressors operate at high efficiency because there is no re-expansion of refrigerant (like there is in the clearance space of a reciprocating compressor).

Compressors can be divided into three groups: hermetic, serviceable hermetic (semi-hermetic) and open drive.

A fully welded hermetic compressor contains a motor and a compressor in a single shell. The only way to service components inside a hermetic compressor is to cut open the shell. Hermetic compressors are usually replaced rather than repaired.

Serviceable hermetic (semi-hermetic) compressors are manufactured with the motor and compressor in a shell that is bolted together. This design enables service of motor or compressor sections.

Open drive compressors are designed with the motor and compressor in two separate sections. The drive mechanism can be either a belt or drive shaft. An open drive compressor incorporates a shaft seal to keep refrigerant from escaping to the atmosphere.

Some compressors are fitted with an unloading device to provide variable capacity. Variable capacity prevents undesirably low evaporator pressures, frost accumulation on the evaporator, and compressor short cycling. Short cycling damage occurs when the motor does not have enough time to dissipate the heat generated from the initial start up current. Short cycling can cause permanent damage to the motor windings.

To operate efficiently, air conditioning systems must be properly sized. Oversized equipment tends to run for shorter periods (short cycle). This can shorten the life of the compressor and cause insufficient dehumidification of the conditioned space.

Condenser

A condenser is a heat exchange unit somewhat like an evaporator, but rather than absorb heat, the condenser removes heat from the refrigerant vapor causing it to condense back to liquid. The first function of a condenser is de-superheating the refrigerant vapor so that condensing can occur. As the refrigerant cools, it condenses, becoming liquid. Air is a non-condensable gas. Any air in the refrigeration system will eventually reach the condenser and remain there, taking up valuable space causing an increase in discharge pressure and a reduction in efficiency. The final pass of refrigerant through the condenser causes the liquid refrigerant to become slightly subcooled. The subcooled liquid refrigerant is now ready to be sent to the evaporator where the cycle begins again. There are three basic condenser types: air-cooled, water-cooled, and evaporative.

Water-cooled condensers condense refrigerant at about 105˚F when the water is supplied at 85˚F with a leaving temperature of 95˚F, resulting in a 10˚F approach temperature. Water-cooled condensers come in many styles, the most common of which are tube within a tube, shell and coil, and shell and tube. Shell and tube is the most expensive style, but the advantage of using it is that it can be opened for cleaning.

An air-cooled condenser condenses refrigerant at 20˚F to 30˚F above entering air temperature, depending on the amount of surface area per ton of condenser. For example, an R-410A air-cooled condenser would have a head pressure of 390 to 445 psig on a 95˚F day. So:

(95˚F+20˚F=115˚F) =390 psig
(95˚F+30˚F = 125˚F)=445 psig

Air requirements for air-cooled condensers are based on the physical size of the system's coil. Air movement can be the same or different for a model design of different tonnage.

Some commercial air conditioning condensers operate during periods of low ambient temperature. Fan cycling, air shutters or dampers, or condenser flooding can be used to maintain a workable head (discharge) pressure. Cleanliness and adequate airflow are crucial to the proper operation of a condenser.

Water-cooled condensers are often used with cooling towers. Water flow through the condenser is controlled by a water regulating valve. The cooling tower re-circulates water, cooling it to within 7° F of the wet bulb temperature. This can add up to substantial savings when compared to a wastewater system.

In an evaporative condenser, the refrigerant coil is located in the tower. The capacity of an evaporative condenser depends on the wet bulb temperature of the entering air.

Metering Devices

Every refrigeration system needs some type of refrigerant control device. Some systems use an expansion valve, while others use a capillary tube. Refrigerant control devices, like compressors, separate the high pressure side of the system from the low pressure side. On the inlet side of the control device there is high pressure/high temperature liquid. The control device causes a drop in pressure, changing the refrigerant to a low pressure, low temperature saturated liquid.

The control device also maintains the correct amount of refrigerant in the evaporator. Too much refrigerant allows liquid to reach the compressor which can cause serious damage. Too little refrigerant prevents the system from working efficiently.

Expansion devices come in the following types: high side float, low side float (flooded evaporator), thermostatic expansion valve, automatic expansion valve, and fixed bore (capillary tube or fixed orifice). **Note:** High-side float and low-side float are seldom used and will not usually be encountered in the field.

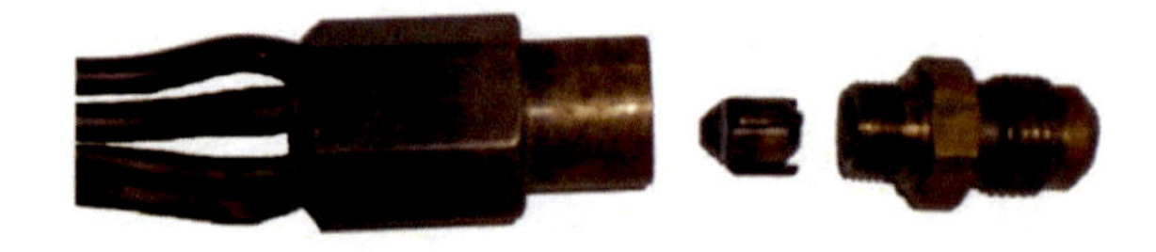

Fixed-Orfice Metering Device

Capillary Tube

A capillary tube is the simplest refrigerant control device consisting of a copper tube with a small inside diameter.

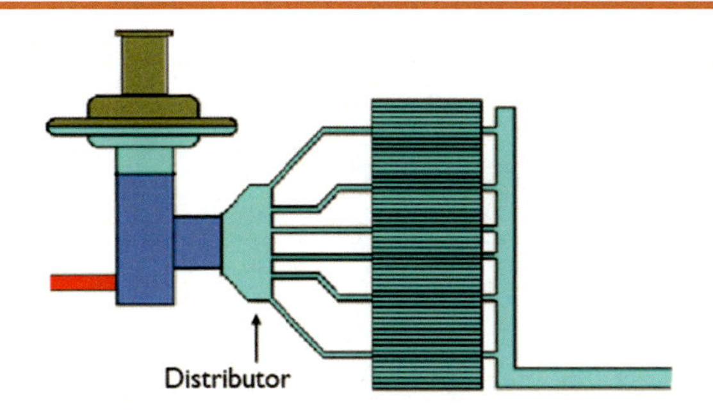

Automatic Expansion Valve

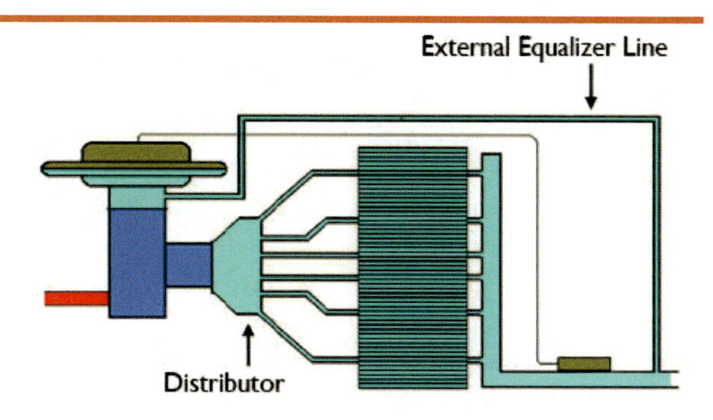

Thermostatic Expansion Valve

The length and diameter of the tube determines the refrigerant flow and pressure drop. Since capillary tubes have no moving parts and do not require adjusting, a properly charged system should operate properly for many years.

A capillary tube is a fixed metering device so the system cannot adjust quickly to load changes. This makes capillary tubes best suited for use in applications where the load remains fairly constant. During the off cycle, a capillary tube continues passing refrigerant until the high and low side pressures have balanced.

Because the inside diameter of a capillary tube is very small, equipment manufacturers usually place a strainer or drier in the line just before the tube to prevent debris from blocking refrigerant flow.

One advantage to using a fixed-bore metering device is that it allows system pressures to equalize during the off cycle. This reduces the starting torque requirements of the compressor motor.

Capillary tube systems are classified as critically charged systems. Refrigerant charge is analyzed when a unit is operating at design conditions. When a unit is operating with the proper amount of refrigerant, a superheat of about 5 to 30 degrees can occur, depending on the application. Some small appliance manufacturers take advantage of heat exchange by soldering the capillary tube along the length of the suction line. In this case, a superheat reading must be taken at the evaporator outlet before its point of contact with the capillary tube.

Capillary tube or fixed orifice systems are slow to respond to changes in load and charge adjustments. The system will take at least fifteen minutes to reach a balanced condition after a charge adjustment is made. Most manufacturers do not recommend adding refrigerant to top off the charge, but they do recommend starting an empty and evacuated system, and then measuring a complete charge into the system.

Automatic Expansion Valve (AEV or AXV)

In some older models of room air conditioners, an automatic expansion valve (AEV or AXV) was used to meter refrigerant. AEVs have the advantage of being able to control the pressure, and therefore temperature, of an evaporator. Maintaining constant pressure can prevent coil freezing. AEVs are best used when the system load remains fairly constant.

Thermostatic Expansion Valve (TEV or TXV)

A thermostatic expansion valve (TEV or TXV) is one of the most commonly used metering devices in air conditioning and refrigeration applications. Thermostatic expansion valves make it possible to keep an evaporator functional under any load condition; because of this, they can be used in a wide variety of applications. TEVs are often referred to as constant superheat valves because they control the amount of liquid refrigerant entering the evaporator by maintaining a constant degree of superheated vapor leaving the evaporator.

Three operating pressures govern the operation of a thermostatic expansion valve: spring pressure, evaporator pressure, and sensor bulb pressure. Evaporator pressure works under the valve diaphragm as a closing force. Spring pressure is also a closing force, transmitted to the underside of the valve diaphragm by means of a pin carrier and push rods. Some TEV designs have a spring pressure that is factory set and non-adjustable, while others provide for field adjustment. When making superheat adjustments, technicians should allow approximately 10 to 15 minutes after each adjustment for the system to reach a balanced condition before another adjustment is made.

To sense variation of superheat in the suction vapor leaving the evaporator, a sensor bulb is attached in close thermal contact to the suction line, as close as possible to the evaporator outlet. The sensor bulb assumes the same temperature as the suction vapor in the evaporator at the point of bulb contact. Bulb pressure is transmitted to the top of the valve diaphragm through a capillary tube. As evaporator superheat increases, the corresponding increase in bulb pressure works against evaporator and spring pressures as an opening force. The increased refrigerant flow lowers evaporator superheat to the set point. Several valve and bulb charge designs have been developed to provide maximum valve performance for each application. Some large evaporators have multiple circuits and a distributor to equally distribute refrigerant. The pressure drop caused by a large evaporator requires a TEV that uses an external equalizer port. The external equalizer ensures that the pressure acting on the bottom side of the TEV diaphragm is the same as the pressure at the evaporator outlet.

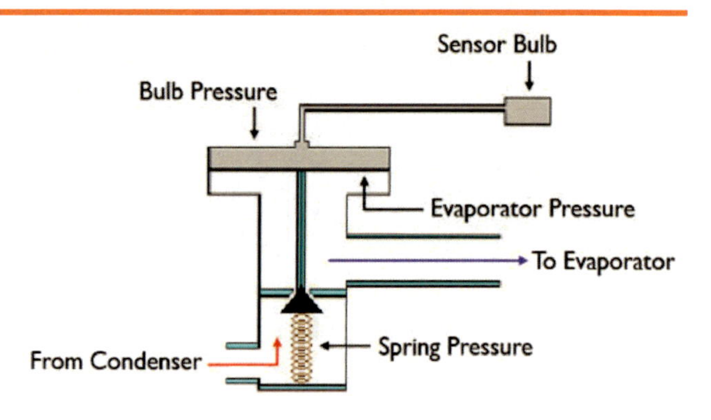

Thermostatic Expansion Valve

Heat Pump

As stated earlier, refrigeration is the process of moving heat. In contrast, a heat pump is a reverse cycle refrigeration system. A special four-way reversing valve is the heart of the heat pump system. This valve allows the evaporator coil and condenser coil to exchange functions. In cooling mode, a heat pump functions as a normal air conditioning system—absorbing heat from indoors and rejecting it outdoors. In heating mode, the coils reverse their functions. The outdoor coil becomes the evaporator and the indoor coil becomes the condenser. Now the heat pump is moving heat from outside the structure to inside.

Heat pump systems can have various designs. Some heat pumps absorb heat from the outdoor air. These models work efficiently in more moderate climates. Other heat pumps work by absorbing heat from either a water well, a coil buried deep in the ground or a coil submerged in a nearby lake or pond.

New technology has led to an increase in the popularity of heat pump systems. Although their efficiency rating continues to increase, most heat pump systems still require some type of supplemental heating system.

Refrigerant Blends

Azeotropic blends are mixtures of two or more refrigerants that behave like a single component refrigerant when evaporating or condensing. A true azeotropic blend has only one boiling temperature and one condensing temperature at a given pressure. Once blended, azeotropic mixtures do not separate.

A **near-azeotropic blend** is a mixture of two or more refrigerants that can still be separated back into the individual components. This separation is called fractionation. Because the composition of these blends varies, so do their boiling (bubble) and condensing (dew) points. This difference is called the **temperature glide.**

Zeotropic blends can also be fractionated but they have a much greater temperature glide than near-azeotropic blends. Zeotropic and near-azeotropic blends should always be charged as liquid.

Compression Ratio

Compression ratio is an important factor when choosing a refrigerant for a particular application. Compression ratio is calculated by dividing absolute suction pressure (psia) by absolute condensing pressure (psia). A single-stage compressor usually requires a compression ratio of 10:1 or lower. Higher compression ratios require the use of a two-stage compressor.

For example:

Discharge pressure + 15 psi = discharge psia
Suction pressure + 15 psi = suction psia
Discharge psia ÷ suction psia = compression ratio

Insufficient airflow or a dirty evaporator negatively effect compression ratio. A dirty or air restricted evaporator does not efficiently absorb heat lowering suction temperature and pressure making the vapor less dense and reducing compressor performance.

Insufficient air flow or poor condenser efficiency effects high-side pressure and the compression ratio. Increased head pressure causes the amount of refrigerant gas in the compressor clearance space to exceed design conditions. Increased discharge temperature accelerates the chemical reactions that form acids and break down oil.

Heat Pump Cycles or Modes

Cooling Mode

During cooling, an indoor coil operates at a low pressure/low temperature and the refrigerant is composed of both liquid and vapor. Refrigerant in the outdoor coil is at a high pressure/high temperature and is both vapor and liquid.

As refrigerant moves through the outdoor coil, heat is removed and the refrigerant becomes a liquid. The removal of heat after this point is called subcooling. During the cooling cycle, the indoor coil serves as the system evaporator and the outdoor coil serves as the condenser. The main differences between a heat pump and a regular air conditioner are the heating and defrost modes.

Heating Mode

When a heat pump is in heating mode, the direction of refrigerant flow is changed by the reversing valve allowing refrigerant to flow from the compressor to the reversing valve and indoor coil.

During heating mode, the indoor coil operates as the condenser. Refrigerant flows from the indoor coil to the metering device feeding liquid to the outdoor coil which serves as the evaporator. At this point, the refrigerant in the outdoor coil is low pressure/low temperature liquid. As the refrigerant evaporates in the outdoor coil, the coil's temperature becomes lower than the ambient air temperature.

Because heat travels from a warmer surface or object to a colder one heat from the air is absorbed. Outdoor air is drawn through and over the cold outdoor coils by the outdoor fan, and the low temperature refrigerant evaporates. The superheated vapor then travels back to the compressor to be compressed, and the cycle continues.

Defrost Mode

Defrost mode removes ice and frost from an outdoor coil during the heat cycle. If not removed, ice and frost will block airflow and insulate the outdoor coil, preventing it from absorbing heat.

When a system goes into defrost mode, the reversing valve changes to the cooling position, the outdoor fan de-energizes and the resistance electric heater(s) energize to provide some heat to the structure.

When ice and frost buildup has been cleared, the defrost control returns the system to normal heating mode. This is the normal cycle for air-to-air heat pumps.

Heat Pump Systems

Important terms:

Supply air: the air flowing into the conditioned space

Return air: the air flowing from the return across the indoor coil

Single-Capacity Systems

Air-to-Air System

In mild climates, the most common heat pump system is air-to-air. In some regions of the United States, winter temperatures rarely drop below 10°F. These regions are where air-to-air heat pumps are the most desirable for many homes and businesses.

Heat pumps may be split or package units, depending upon the type of structure they are designed to heat. For example, a house built on a slab would probably have a split unit with ductwork installed in the attic and the air hander in the garage or an interior closet space.

Split System

A split system's compressor, reversing valve, heating metering device and refrigerant coil are located outdoors.

The air handler inside the structure contains the indoor refrigerant coil, cooling mode metering device, blower assembly and auxiliary heaters. This system is very versatile because the air handler can be mounted in a basement, indoor closet or garage, on a rooftop or even in an attic.

Air-to-Air Package Unit, or Unitary Unit

An air-to-air package unit, also called a unitary unit, houses all system components in one package and is usually installed outdoors. The unit includes the housing, indoor and outdoor coils, metering devices, reversing valve, fan and blower assemblies, auxiliary heaters, compressor, and electrical system controls. The indoor thermostat is mounted on a wall inside the structure.

Package Unit

A package unit does not require a technician or installer to attach refrigerant lines or charge the system with refrigerant. The package unit usually comes "service ready" from the manufacturer except for the electrical power supply, thermostat wiring, thermostat, and duct system.

Geothermal Systems: Water-to-Air System

Water-to-air systems are used in residential and commercial markets, usually as geothermal heat pumps. The water sources can be from wells, rivers, lakes, ground water or municipality water supply.

In very cold weather, water (geothermal) systems are more efficient than air-to-air systems. Water changes temperature very slowly, whereas outdoor air temperatures can change drastically in short periods of time. Water-to-air systems can produce a coefficient of performance (COP) of 4:1 or more during the heating season.

During the cooling season, outdoor air temperatures can soar to over 100°F, while lake water temperatures usually do not rise above the low nineties. This means that a water-to-air system will perform more efficiently in very hot weather.

Section 1: Basic Principles of Operation

A vertical or horizontal closed-loop system can supply heat for a structure from a heat exchanger filled with municipality water and, unlike air-to-air systems, does not require a resistance heater for auxiliary heat.

Water-to-Water System

Water-to-water systems are used mainly in commercial applications, but can also be used to heat water for domestic use and to supply hot water to the coils in air handlers.

Ground-to-Air System, or Direct Coupled System

Ground-to-air systems are sometimes used in residential and commercial markets. In a ground-to-air system, the outdoor coil or refrigerant circuit is buried in the ground.

Operating costs are generally low; the ground is an extremely good source for heat in the winter and a good place for heat rejection in the summer.

Single-Speed Compressor

Most single-speed compressors used for residential and light commercial heat pump applications are PSC (Permanent Split Capacitor) compressor motors. These compressors use only a run capacitor for starting and running.

Some manufacturers use a positive temperature coefficient (PTC) thermistor with these compressors. The PTC bypasses the run capacitor while the compressor is starting, but is out of the electric circuit while the compressor is running. These devices are used for soft-starting purposes only.

A CSR (Capacitor Start Capacitor Run) compressor is recommended when a system has a thermostatic expansion valve metering device. Most TXV systems do not allow the pressures to equalize during the off cycle. If the low and high side pressures are not equalized on start-up, the compressor will overload and cycle on and off.

Many manufacturers offer a start kit that can be added to the PSC compressor. This kit contains a potential relay or solid state PTC device and start capacitor which is used to help the compressor start. A PSC compressor with a start kit is essentially operating as a CSR compressor.

Dual Capacity and Two-Speed Systems

Most heat pump compressors are single speed/single capacity but in recent years, some major manufacturers have developed very innovative changes to improve both efficiency and performance.

Dual Capacity Compressor

Dual capacity compressors use a method that involves reversing the compressor and floating one of the pistons.

Two-Speed Compressor

A two-speed compressor has the advantage of being able to change from high-speed to low-speed operation. This is done by changing the number of poles in the compressor. The more poles the compressor has the slower it can start and run.

Section 2: Compressors, contains a more in-depth discussion of dual-capacity and two-speed compressors.

Review Questions
Section 1: Basic Principles of Operation

Name: ________________________________ **Date:** ________________

1. What unit of measurement indicates pressures below atmospheric?

2. Atmospheric pressure is _________ psia at sea level.

3. Refrigeration is defined as the movement of _________ from an area where it is not wanted, to an area where it is less objectionable.

4. As a refrigerant absorbs heat, it changes from a _________to a ________.

5. Superheated vapor is a gas that has been heated to a temperature _________ its saturation point.

6. During the heating cycle, the outdoor coil of a heat pump becomes the _________.

7. During the cooling cycle, the indoor coil of a heat pump becomes the __________.

8. When a heat pump system goes into defrost mode, what process takes place so that system can continue to provide heat to the structure?

9. A measurement of the speed or motion of molecules in a substance is called ___________.

10. Name two common temperature scales used to measure air temperature.

11. What does ambient mean?

12. What is a wet bulb measurement used for?

13. What is dew point?

14. What is pressure?

15. What does PSIG stand for?

16. The three methods of heat transfer are __________, ____________ and __________.

17. What does a pressure temperature chart reveal?

18. Heat always travels from ________ to __________.

19. By what method does refrigerant release heat?

20. What is the pressure of refrigerant 410A at 30F°?

21. What is the pressure of refrigerant 410A at 120F°?

22. In order, list the three processes refrigerant goes through in the condenser:

23. In order, list the two processes refrigerant goes through in the evaporator:

24. What are the differences between a heat pump and regular air conditioner?

Lab Assignment 1

Section 1: Basic Principles of Operation

Name: ______________________________ **Date:** ____________________

Objective: Familiarization of available lab equipment.

Directions: Identify lab heat pump equipment types, capacity, and general features. Fill in the information and check boxes that apply to the equipment used for training.

System 1

Type of System

Brand: ________________________

☐ Window unit ☐ Package

☐ Split System ☐ PTAC

Model Number: ____________________

☐ Mini Split ☐ Other

Serial Number: ____________________________________ Voltage: __________

Capacity: ________________ Btu Ampacity: __________ Refrigerant Type: ________

System 2

Type of System

Brand: __________________________

☐ Window unit ☐ Package

☐ Split System ☐ PTAC

Model Number: ____________________

☐ Mini Split ☐ Other

Serial Number: ____________________________________ Voltage: __________

Capacity: ________________ Btu Ampacity: __________ Refrigerant Type: ________

System 3

Type of System

Brand: ________________________

☐ Window unit ☐ Package

☐ Split System ☐ PTAC

Model Number: ____________________

☐ Mini Split ☐ Other

Serial Number: ____________________________________ Voltage: __________

Capacity: ________________ Btu Ampacity: __________ Refrigerant Type: ________

Objectives

Upon completion of this section, the participant will be able to:

1. name the characteristics and purpose of different compressors used in heat pump systems;
2. identify the use, components and operating characteristics of fan and blower motors;
3. explain the operation of valves and metering devices;
4. identify various components including check valves, filter driers, accumulator and defrost controls.

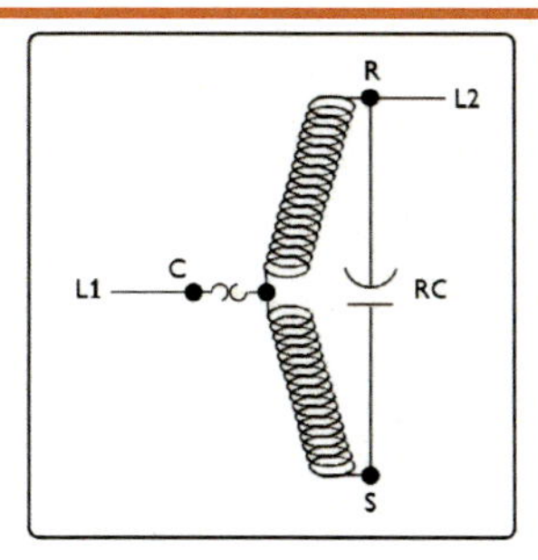

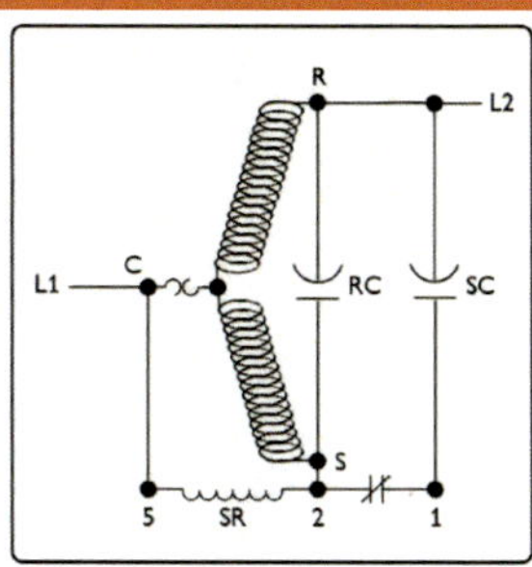

Single-Speed Compressor

Compressors

The compressor is the heart of the system. It creates a pressure difference between the low and high sides of the system so refrigerant flows.

The volume the compressor pumps is one factor in determining overall system capacity. Physical size of the coils, refrigerant type used, metering device, and system airflow are important in determining total system capacity.

The compressor must be able to operate at high compression ratios due to various outdoor temperatures. The system operates as a high temperature air conditioner during the cooling season and as a low temperature refrigeration system during the heating season.

Various Types of Compressors Used With Heat Pumps

Reciprocating: Piston driven compressors have been used in the refrigeration industry for years. Energy costs, noise pollution, and government mandated energy standards required development of new compressor technologies. To meet these needs and to keep manufacturing costs low, rotary and scroll compressors were developed.

Rotary: Rotary compressors were used in split systems for a while, but are now primarily used in PTACs and window units. The rotary compressor is lightweight, durable, and efficient. Liquid flooding can damage rotary type compressors if the system is over charged.

Scroll: Scroll compressor theory has been known for many years, but the high cost of production and manufacturing difficulties delayed production until development of new computer numerical controlled (CNC) machine tools. CNCs made it possible to manufacture scroll compressors to the close tolerances required for proper operation. These compressors can be found in many popular units. The scroll compressor has no pistons-only *fixed* and *moving* spiral mechanisms inside.

Since the introduction of heat pumps, the reciprocating compressor was the manufacturer's choice due to its dependability and performance capabilities. The permanent split-phase (PSC) compressor was used in both residential and industrial heat pump systems. Most manufacturers used the "hermetic compressor" and refrigerant R22 in their residential heat pump systems. In the past, "semi-hermetic" compressors with other refrigerants were used. Todays, technology brought the Scroll compressor into use.

As shown above, many single phase heat pumps are equipped with either a PSC or a Capacitor Start/Run (CSR) compressor motor. The CSR motor has both a run capacitor and start capacitor to start the compressor. The start capacitor must drop out of the circuit after compressor start. The PSC motor uses a run capacitor which stays in the circuit. These motors have an internal overload device embedded in the windings which open if the temperature of the windings reaches a preset point. The overload switch breaks the common terminal connection to the run and start windings. It automatically resets after the compressor motor has cooled down to a preset temperature.

Variable Speed Compressors

One of the first variable speed systems developed for residential use featured a three-phase compressor with a frequency drive. The next generation consisted of a three-phase compressor with permanent magnets mounted on the rotor, and a speed drive to supply electrical power.

In commercial applications, variable frequency drives are used with AC motors. Variable speed drives are used for DC and AC motors. A variable speed drive modulates the voltage level and time it is applied to the motor. A variable frequency drive lowers the frequency below 60 cycles to slow the motor and raises it above 60 cycles to speed up the motor.

Normal RPM is based on 60 Hz
Variable RPM: 0 Hz to 90 Hz equals 0 to 10,800 rpm

Variable-Speed Drives (VSD)

Variable frequency drives (VFD)

Variable frequency drive motors operate on the same basic

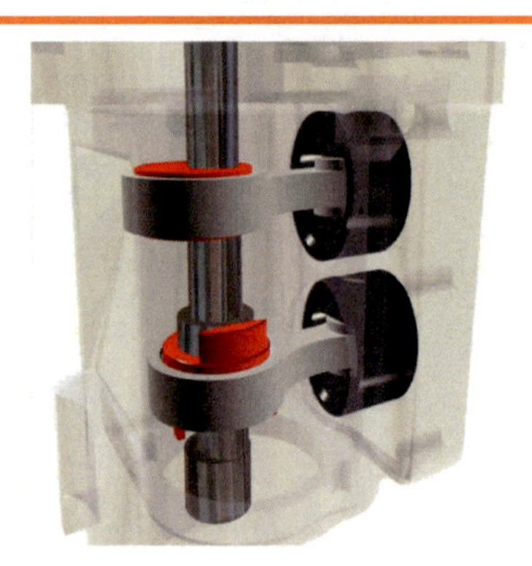

Bristol TS Compressor

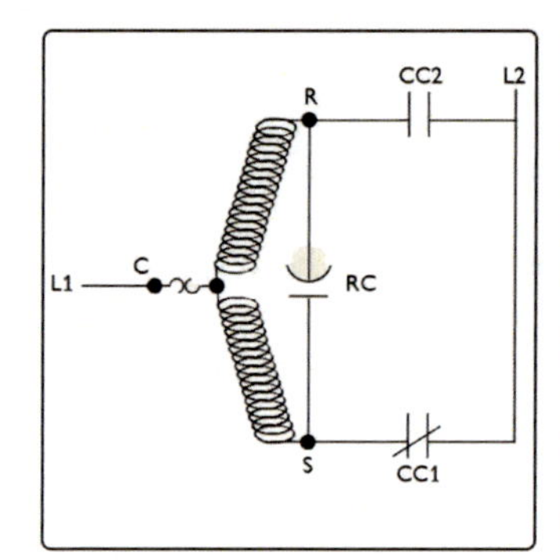

Low-Capacity Wiring

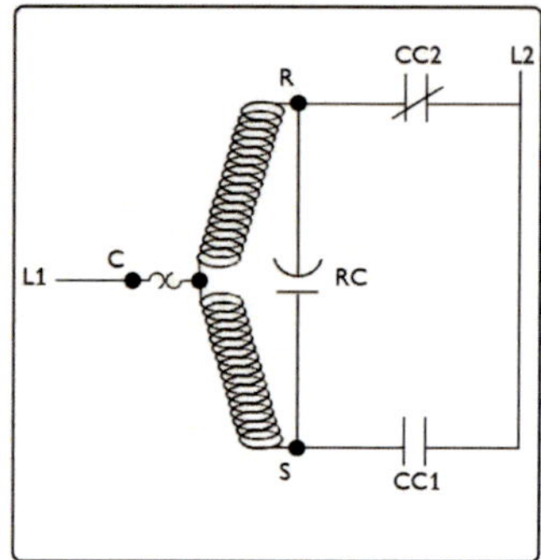

High-Capacity Wiring

principles as synchronous AC motors. The number of poles in the stator winding and frequency determine the speed.

Dual Capacity Reciprocating Compressor

Also known as the TS or Twin/Single piston compressor, this compressor is fairly new to the residential market. The compressor is manufactured with a dual eccentric shaft and two pistons. One of the eccentric rings is solid or fixed, and the other is a two-piece design. When wired as a normal PSC motor, the compressor uses both pistons. When the compressor runs in reverse, only one piston operates because one of the eccentric rings changes position. The second eccentric consists of two separate rings. When the outside ring rotates a half turn to the stop, it centers itself to the main shaft and the piston no longer moves back and forth.

Reversed Run Dual Capacity Reciprocating Compressor

For low capacity operation, the contactor (CC1) connected to the start winding is energized. By applying line voltage to the start winding instead of the run winding, the PSC compressor runs in reverse. Running the compressor in reverse makes the bottom eccentric return to the center position, which prevents the piston from pumping. Other commonly used compressors would not run if wired this way. This compressor runs because only one piston is pumping and the load of the compressor motor is reduced by one-half. This load reduction also reduces the current flow through the windings by approximately one-half.

The Two-Speed Compressor in High Speed Operation

For *High Speed Operation*, the compressor is wired to the run winding, which has two poles wired in parallel, and energized with a two-pole start winding. (See figure at top of next page.) The run capacitor is in series with the two-pole start winding. This application can be used for reciprocating, scroll and rotary compressors.

Two-Speed Compressor, Low Speed

The diagram (see top of next page) shows the dual speed heat pump compressor wired for operating in low speed mode. Low speed start windings consist of four poles, and are energized with the run winding (two poles in series). The number of poles in the start and the run windings determines speed of the compressor. Low speed operation allows the heat pump to operate at approximately one-half of capacity.

Fan/Blower Motors

Condenser Fan Motor

With residential heat pumps, most outdoor fan motors are *Permanent Split Capacitor (PSC).* It is a favorite for air-handlers. These motors are often multi-speed, 1/2" to 5/8" shaft, sleeve-type bearings and usually range from 1/6 to 3/4 horsepower, depending on application. They are durable, dependable, and less costly, with very low maintenance. Most are factory lubricated with oil and sealed. The condenser fan motor is usually designed for direct drive applications to turn a fan blade matched to the CFM requirements of the manufacturer. It must be waterproof since it is subjected to weather. The synchronous (calculated) speeds for these motors are 900, 1200, 1800, and 3600 RPM. Due to magnetic slippage of the rotor, the approximate speeds of these motors are 850, 1140, 1725 and 3450 RPM respectively. The difference between synchronous (calculated) speed and actual speed is "*slippage*". This speed difference is caused by the load or work being done.

To determine the synchronous speed of PSC motors:

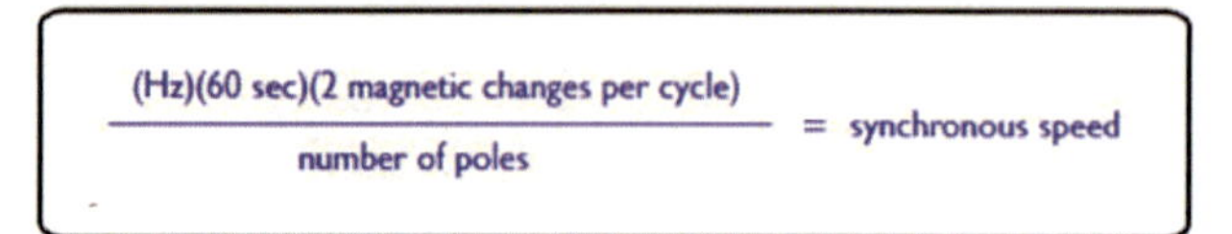

(Hz)(60 sec)(2 magnetic changes per cycle)
number of poles
= synchronous speed

The sleeve bearings are made of bronze, porous iron or steel-backed Babbitt and have oil grooves and wicking for lubrication. These motors can be mounted horizontally or vertically. A plastic water-repelling shield is often placed over the shaft of the motor to help keep water and moisture from entering the bearing when mounted vertically.

Some fan motors have *ball bearings* inside a race mounted in each of the *end bells.* These motor bearings are *never oiled.* In smaller horsepower motors *ball bearings are packed with grease* and sealed by the manufacturer.

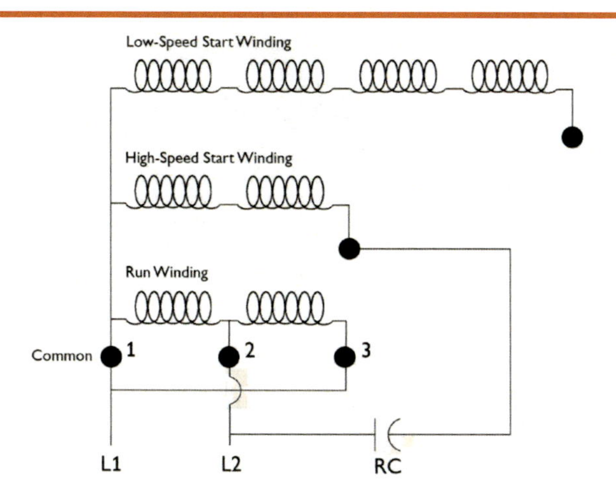

Two-Speed Compressor in High-Speed Operation

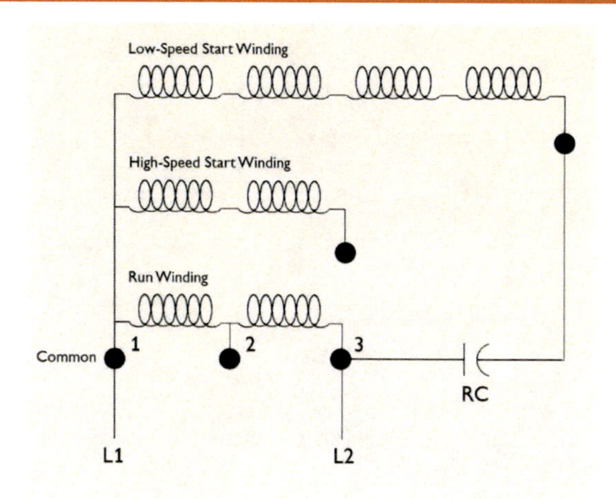

Two-Speed Compressor in Low-Speed Operation

Fan motor mounting methods:

1. four studs holding the motor together
2. flange mount
3. frame mount
4. resilient mount

The most important part of the motor is the *Name Plate.*

The **name plate** tells horsepower, RPM, specified voltage, phase and Hertz, amperage draw, frame number, duty rating, type, thermal protection type, wiring information, lubrication (if needed) and motor rotation.

The Indoor Blower Motor

Air handler blower motors are usually Permanent Split Capacitor or variable speed which varies speed to meet manufacturer CFM requirement.

The Permanent Split Capacitor motor has two major parts: STATOR and ROTOR.

The Stator is the stationary part which houses the motor windings and stacked laminations.

The Rotor consists of the shaft, rotor core and an internal fan for cooling the motor.

End Bells are the end covers housing the bearings. The bearings must hold the rotor in the exact center position inside the stator.

The variable speed motor or Electronically Commutated Motor (ECM) has factory preset speeds inserted into its digital logic control board and dip switches (or jumpers) for speed or CFM selection.

4-way/Reversing Valve

The four-way (or reversing) valve directs flow of refrigerant to either the indoor or the outdoor coil. An electrically operated pilot valve controls pressure on each side of the main slide valve (pilot valve may have 3 or 4 connections). There must be a minimum pressure difference of 75 PSI between the low and high sides to make the main slide valve move. Under normal operating conditions, only refrigerant vapor flows through the reversing valve. The refrigerant lines in the center of the valve connect to the compressor. The smaller line, located in the middle of the four-way valve, connects to the compressor discharge line. The larger center line of the three on the other side of the four-way valve connects to the inlet side of the suction accumulator (if used), or to the compressor suction line. The function of the two lines never changes. The two outside lines connect to the indoor and outdoor coils. Which line connects to the outdoor coil is determined by whether the valve is energized in cooling or heating mode. The two outside lines alternate from hot discharge vapor to cool suction vapor depending on mode of operation.

Indoor Blower Motor

Variable-Speed Indoor Blower Motor

Four-Way Reversing Valve

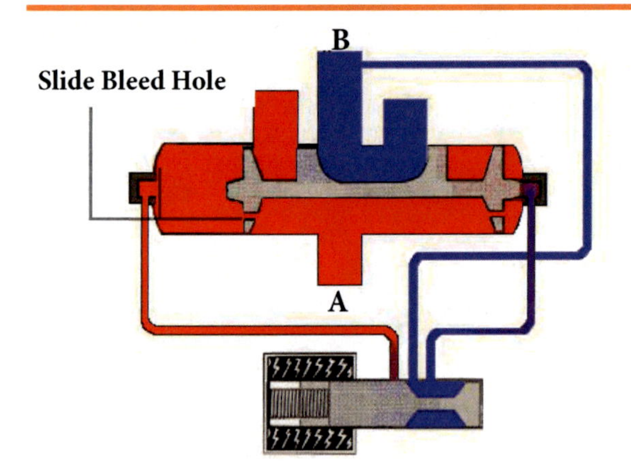

Four-Way Reversing Valve, De-Energized

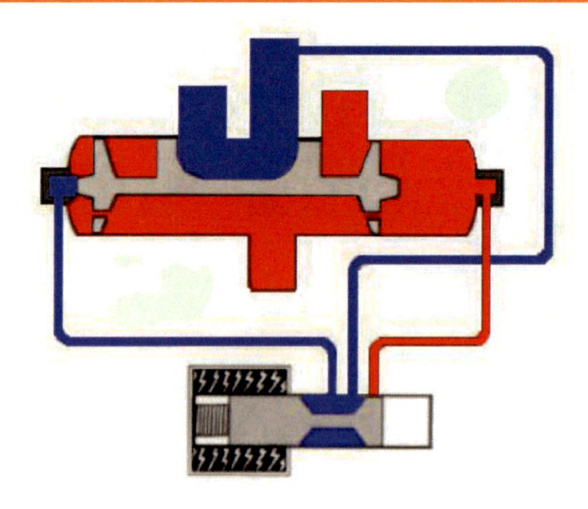

Four-Way Reversing Valve, Energized

Reversing Valve Operation-Solenoid De-Energized

In the example above, the pilot-valve solenoid is de-energized allowing refrigerant on the right side of the main slide to be at the same pressure as in the suction line. The center and right capillary tube connections on the pilot valve are open to each other allowing high pressure on the right side of the main slide to flow into the suction line. The pressures on the left side and middle of the slide is high pressure, which cannot bleed through the left capillary line of the pilot valve. This keeps the slide in the right position, making the right top line *suction* and the top left line *discharge.*

Solenoid Energized

When the pilot-valve solenoid is energized, the pilot valve slide moves to the left, opening the left capillary line to suction and blocking the right capillary line. The high-pressure refrigerant on the left side of the main slide is allowed to flow into the low side suction, which drops pressure on the left side of the main slide. The bleed **orifice** in the main slide piston is smaller than the bleed lines of the pilot valve. This allows pressure to drop faster than it can build up, creating a pressure difference across the main slide valve and making it move to the lower pressure. When the main slide moves all the way to the left, the bleed port to the pilot valve is sealed off by the valve seat, which is located on the end of the main slide. When the pilot-valve solenoid is de-energized, pressure is reduced on the right side, and the process is repeated in reverse. Some pilot valves force the main slide to move by directing high pressure to one side of the valve or the other. These pilot valves have four connecting capillary lines instead of three.

Metering Devices

The metering device is the point in the system where high pressure liquid refrigerant is metered and changed into low pressure/low temperature liquid and fed into the evaporator coil.

Many metering devices are used for heat pumps including:

- Flow and Bi-flow control devices
- Thermostatic and electronic expansion valves
- Capillary tube

The primary purpose of a metering device is to control flow of liquid refrigerant into the evaporator. It does this by feeding liquid refrigerant from the high-pressure liquid line into the evaporator, which is operating at a much lower pressure and temperature. As refrigerant flows through the metering device, a small percentage of the refrigerant flashes cooling the remaining refrigerant to the operating evaporator temperature. The size of metering device needed is determined by the Btu capacity, the pressure drop across the valve, and the type of refrigerant used. The type of metering device has an effect on the efficiency of the system.

A heat pump can be designed to operate with the same or different types of metering devices on the indoor and outdoor coils. The system may contain capillary tubes, one set for heating and one set for the cooling cycle or one set of capillary tubes and one TXV. Some manufacturers are using one bi-flow valve for heating and one TXV for the cooling cycle.

Capillary tube metering devices are normally used with a ball-type check valve, which is piped in parallel with the metering device. The fixed orifice metering device is usually made to meter flow in one direction and allow refrigerant to bypass around the orifice when the direction is reversed.

Three types of thermostatic expansion valves commonly used:

1. TXV—allows refrigerant to flow in one direction only, and has a mechanical check valve piped in parallel for reverse flow bypass.
2. Bi-flow TXV—has a bypass valve built into the valve body.

3. Electronic controlled TXV—used in some commercial applications and are modulated open for reverse flow bypass.

ExpansionValve

Check Valves

Check valves allow fluid flow in one direction only. In a heat pump system, check valves are used to bypass refrigerant flow around the metering device not in use. Refrigerant flowing to a check valve either opens or closes the valve, depending on the direction refrigerant is flowing when approaching the valve. A heat pump system with two metering devices uses only one check valve for each mode of operation (cooling/ heating).

In cooling mode, the outdoor coil check valve bypasses refrigerant flow around the *heating* metering device and the indoor check valve blocks refrigerant flow from bypassing the *cooling* metering device.

In heating mode, the outdoor coil check valve blocks the flow of refrigerant from bypassing the heating metering device and the indoor check valve bypasses refrigerant around the cooling metering device.

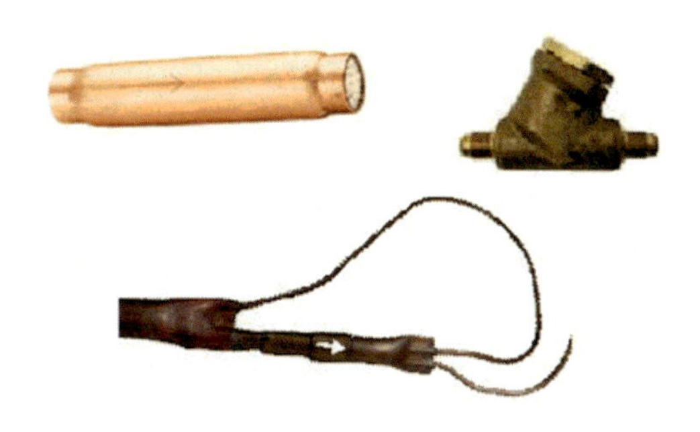

Check Valves

On some package or window unit heat pumps, only one check valve is used. A short capillary tube in series with the main capillary tube is bypassed for the cooling cycle. For the heating cycle, refrigerant flows through both capillary tubes. An arrow on the check valve body indicates the direction of refrigerant flow.

Fixed-Orfice Metering Device

The fixed orifice piston works as a metering device serving the coil it is connected to and as an open check valve, when refrigerant flow is from the coil. This combination check valve and metering device is used on most air-handlers with cooling and heat pump applications.

Liquid line filter drier *Suction line filter drier*

Filter Driers

A filter drier is designed to absorb moisture and capture foreign matter present in a system. There are several locations where manufacturers install liquid line filter driers on heat pumps. One filter drier may be used in the outdoor unit, which filters only on the cooling cycle. Two filter-driers may be used - one in series with each check valve. The liquid line filter drier must be placed in the liquid line, in the right direction. The direction of flow in the liquid line changes depending on whether the system is operating in cooling or heating mode.

Some liquid line filter driers for heat pumps are different from those used on straight air conditioning units. Standard driers filter in one direction only. On a heat pump the drier is placed in parallel with one of the check valves to maintain the direction of flow. The bi-flow drier contains internal check valves that allow refrigerant to flow in either direction through the drier. This filter drier can be installed in the main liquid line. The use of the directional flow or the bi-flow filter drier depends on where it is placed in the system.

The drying agents (desiccants) inside the driers consist of:

- Activated alumina (formed from aluminum oxide).
- Silica gel (polymerized silica).
- Molecular sieve is very effective for moisture.

A suction line filter drier protects a newly installed replacement compressor. If a system has a compressor burn-out, the filter drier should be used to trap acids.

Any drier with a substantial temperature or pressure drop across it while in operation is partially restricted and must be replaced.

Defrost Controls

Mechanical Defrost Control: Some of the first defrost controls consisted of a differential defrost thermostat that sensed the coil and air temperature. Another type was a combination consisting of a defrost thermostat with a timer, or with a static pressure switch.

On larger systems, a combination of controls is used with a coil temperature-sensing thermostat. An air static pressure switch, timer, or both would be used with the thermostat to prevent the system from going into defrost on temperature only. The termination of the defrost cycle is most always performed by the defrost thermostat and is limited by time, if a timer is used as part of the control system.

Mechanical Defrost

Electronic Defrost Control: Electronic defrost controls use the same basic principles as mechanical timer controls. The electronic type provides precise control without moving parts. In the 1980's most manufacturers switched from mechanical defrost controls to electronic defrost controls.

Electronic Defrost Control

Defrost Thermostat: The defrost thermostat is normally a snap-disk type thermostat clipped on the outdoor coil. It prevents the system from going into defrost until the coil temperature is low enough. Most defrost thermostats close below 32°F and open to terminate the defrost cycle above 45°F.

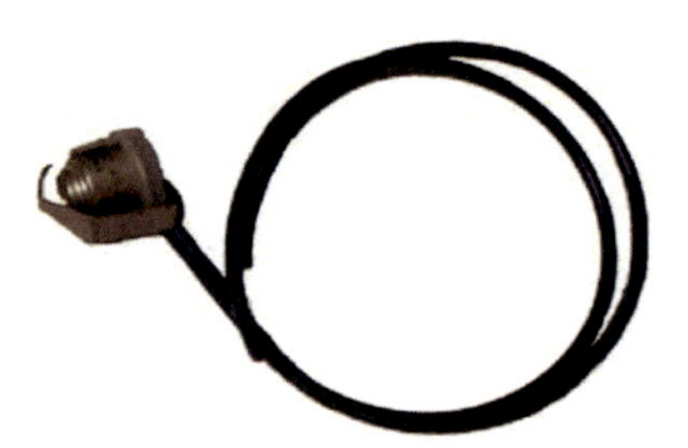

Defrost Thermostat

Defrost Relay: Normally a three-pole relay used with some defrost controls. It controls loads such as the outdoor fan, reversing valve, and first stage of electric heat while the system is in defrost mode.

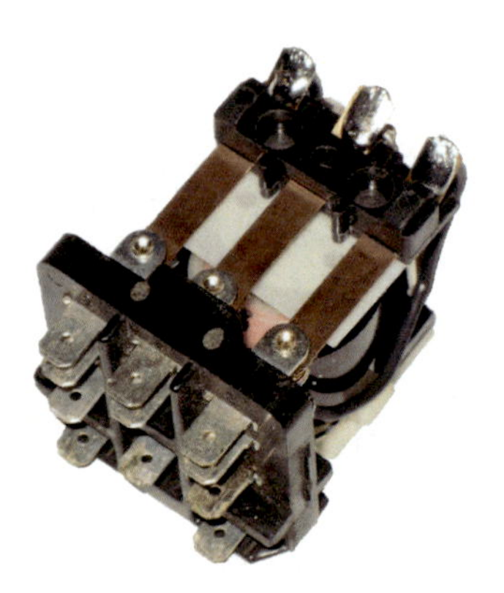

Defrost Relay

Control Thermostats for Heating & Cooling

Types of control thermostats include:

- Single stage manual change-over
- Single stage automatic change-over
- Two stage manual change-over
- Two stage automatic change-over

Control Thermostat

Most split system heat pumps use a single stage cooling and a two-stage heating thermostat. The first stage of heating operates the heat pump system and the second stage of heating controls the supplemental (or auxiliary) heat.

An automatic changeover thermostat has four mercury bulbs used to switch the heat pump to the heating cycle or the cooling cycle as needed. The first *stage of cooling* controls the reversing valve solenoid, and the second stage controls the contactor. The first stage of *heating* controls the contactor and the second stage controls the electric heat.

Control wiring may be a minimum of 5 conductors to more than 10 conductors depending on system and accessories.

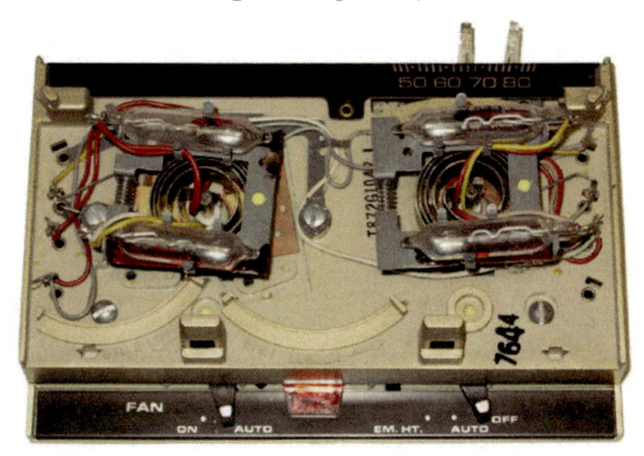

Automatic changeover thermostat

Electronic Control Thermostat

Electronic control thermostats replace mechanical thermostats. The electronic thermostats offer more features reducing inventory for dealers. For homeowners, they provide better control of the system improving efficiency. Some models provide options of multiple setback times, and temperature settings for each day of the week. When used with an ECM blower motor some thermostats can control indoor humidity.

Solid state technology is improving at such a fast rate it is almost impossible to keep up. The electronic thermostats today are mini-computers. In some cases, they can be used to control or monitor total system operations such as electrical, refrigeration cycle, air flow, and system efficiency. If a problem develops, the thermostat can automatically call the dealer for service.

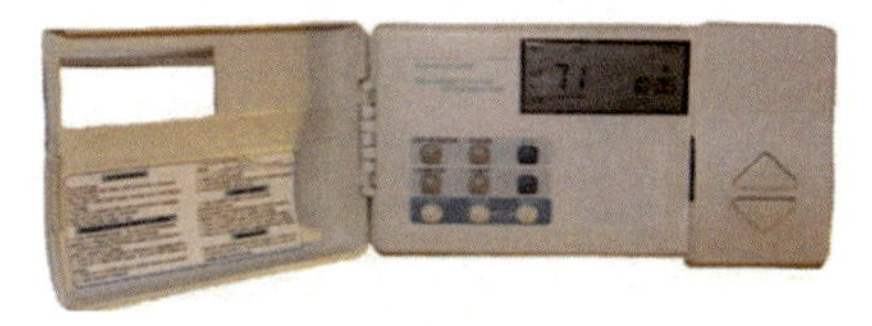

Automatic control thermostat

Crankcase heaters

The main purpose of a crankcase heater is to prevent refrigerant from migrating into the oil. Commonly used crankcase heaters are:

- Wrap around,
- Well type,
- Internal motor winding heat.

The compressor sump should be approximately 20°F warmer than other parts of the system to prevent refrigerant migration. If the compressor is not at least 5°F warmer, the refrigerant can condense in the bottom of the oil sump or in the cylinder, rotor chamber, etc. Liquid refrigerant damages a compressor by washing out the oil or by creating a high compression condition which breaks internal mechanical parts.

A crankcase heater does not prevent liquid slugging from a severe overcharge of refrigerant.

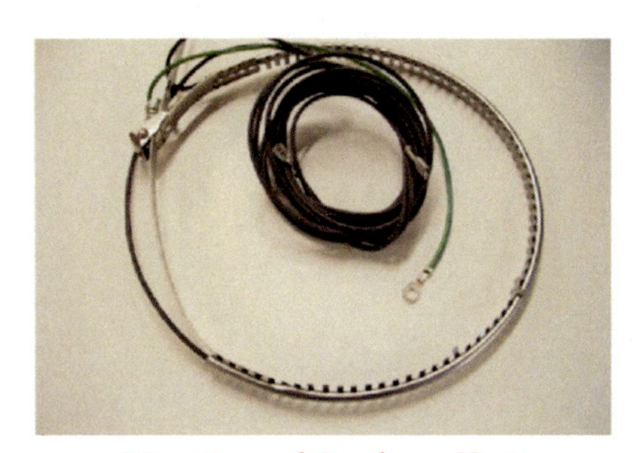

Wrap-Around Crankcase Heater

Suction Line Accumulator

Low temperatures where a heat pump operates in the heating cycle make it necessary to install an accumulator. This protects the compressor from liquid flooding. The accumulator catches/stores liquid refrigerant before it reaches the compressor. It is mounted in the true suction line between the compressor and the reversing valve. When the temperature at the accumulator increases, the liquid boils off and returns as a vapor to the system. The accumulator is typically sized to hold 50% of the total system charge.

Note: Not all manufacturers will use suction line accumulators.

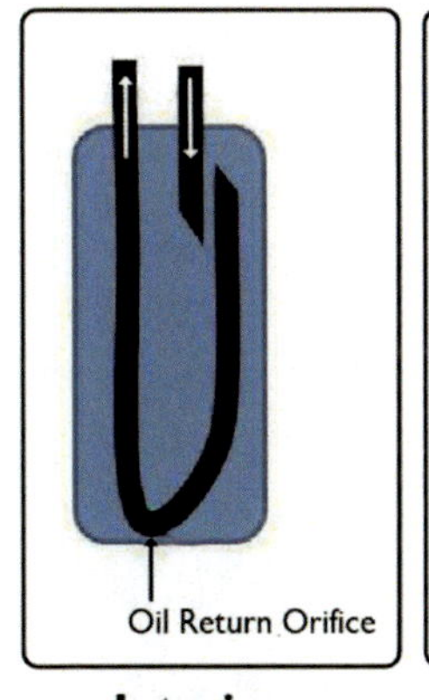

Interior **Exterior**

Suction line accumulator

Review Questions
Section 2: System Components

Name: ______________________________ **Date:** ________________

1. The three most common compressors used in heat pump systems include the __________, __________ and __________.

2. One of the first variable–speed compressor systems developed for residential use featured a three-phase compressor with a ______________________drive.

3. The ________________ prevents the system from going into defrost until the coil temperature is low enough.

4. Which component changes the heat pump refrigeration cycle from one mode to the other?

5. While in the defrost mode the refrigeration system is actually switched into the __________ cycle.

6. What is a check valve and what is its purpose on the heat pump?

7. What makes the switchover valve switch when energized?

8. What type of filter drier is used in the liquid line of a split system heat pump?

9. What does the fixed orifice device do in a heat pump system?

10. A defrost thermostat closes its contacts at approximately ___________ °F, and opens at about_______ °F.

11. List the minimum connections or conductors to the outdoor unit required for a conventional 24 volt control system with electric auxiliary heat.

12. Variable speed Electronically Commutated Fan Motors has what advantages?

13. Most important part of a motor is its ________________ ________________.

14. What is the function of a crankcase heater?

15. Calculate the synchronous speeds for a two, four, and six pole motors.

16. What is an accumulator used for on a refrigeration system?

Lab Assignment 1

Section 2: System Components

Name: ________________________ **Date:** ______________

Objective: Familiarization of compressors type and wiring.

Directions: Identify and list heat pump compressor properties.

NOTE: Perform the following task using all electrical safety procedures!

1. Outdoor Unit Model Number: ____________
2. Compressor Type, check one: ☐ Reciprocating ☐ Scroll ☐ Rotary
3. Measure the resistance for the compressor windings and record the resistance in the space provided in the diagram.

____Ω A ____Ω

C B

____Ω

4. Indicate the appropriate terminal by placing the correct letter from the diagram in the space for each winding.

Run ______ Start _______ Common _____

5. Compressor motor type: □ Induction Start □ PSC □CSR
6. Compressor full load amperage? _____________
7. Compressor locked rotor amperage: _____________
8. Compressor discharge line diameter: _____________
9. Compressor suction line diameter: _____________
10. Run capacitor rating: _________ µf ___________ Vac
11. Start capacitor rating: _________ µf ___________ Vac

Lab Assignment 2

Section 2: System Components

Name: ______________________________ **Date:** __________________

Objective: Familiarization of indoor air delivery systems used for heat pump systems.

Directions: Identify heat pump equipment indoor section air handler/blower, general features.

NOTE: Perform the following task using all electrical safety procedures!

Model: __________________________ Type check one: □ Window unit □ Package

Split System PTAC

1. Number of Speeds: ____________________ □ Mini Split □ Other

2. Motor type check one: □ Induction □ Capacitor Start □ Shaded Pole □ PSC □ ECM

3. Voltage: __________ 4. FLA: __________ 5. Capacitor Rating: ________

6. Fan ____________________horse power 7. Rotation from lead end: □ CW □ CCW

8. Shaft diameter: _________ 9. Shaft length: _________ 10. Blade/wheel diameter: _________

11.Make a sketch of the motor with leads indicating lead colors and identifying the connections.

(Use backside if needed)

Lab Assignment 3

Heat Pumps Section 2: System Components

Name: ______________________________ **Date:** ____________________

Objective: Familiarization with types and combinations of metering devices used on heat pumps.

Directions: Identify heat pump meter devices.

NOTE: Perform all task using all electrical safety procedures!

Indoor Metering Device Type, check one: ☐ Capillary tube ☐ Fixed Orifice ☐ TXV

If capillary tube; How many? __________ How many check valves? __________

If the metering device is a TXV, is the check valve internal? ☐ Yes ☐ No

Outdoor Metering Device Type, check one: ☐ Capillary tube ☐ Fixed Orifice ☐ TXV

If metering device is a capillary tube: How many? ________ How many check valves? ________

If the metering device is a TXV, is the check valve internal? ☐ Yes ☐ No

Lab Assignment 4
Section 2: System Components

Name: ____ **Date:** ______________________

Objective: Gain knowledge and skill in troubleshooting an internal valve leak.

Directions: Identify the type and operation of a reversing valve.

NOTE: Perform the following task using all safety procedures!

Does the pilot valve have 3 or 4 lines? □ 3 □ 4

Is the solenoid coil low or high voltage? □ Low □ High

Is the valve energized for cooling or heating? □ Cooling □ Heating

Use an electronic thermometer to check for a leaking valve on both the cooling cycle and heating cycle.

1. With heat pump system turned off, install electronic temperature probes, one on the center suction line 2 to 3 inches from the valve body. The other probe on the vapor line to the indoor coil 2 to 3 inches from the valve body.
2. Insulate both temperature probes with a rubber type insulation.
3. Operate system for 10 to 15 minutes on the cooling cycle and record the temperature for each line.

   ____________ °F Suction line ____________ °F Vapor/coil line

4. Turn off system.
5. Remove insulation and temperature probe from indoor coil vapor line.
6. Attach the temperature probe the to the outdoor coil vapor line 2 to 3 inches from the valve body and insulate.
7. Operate system for 10 to 15 minutes on the heating cycle and record the temperature for each line.

   ____________ °F Suction line .____________ °F Vapor/coil line

Does the reversing valve change to the heating and cooling positions ? □ Yes □ No

Is the valve operating properly? □ Yes □ No

What could cause the valve not to change position if the pilot valve operates and the temperature check seems to be ok?

____

____

Objectives

Upon completion of this section, the participant will be able to:

1. check and calculate indoor air flow
2. calculate CFM
3. calculate sensible heat ratio
4. determine blower performance

Indoor System Air Flow

Most manufacturers recommend 400 to 450 CFM of indoor airflow to prevent high discharge pressures during the heating cycle. This CFM range works well for heating, but is not always practical for the cooling cycle. CFM for the cooling cycle should be calculated first, using the sensible and latent heat from the structure's heat gain. However, the final CFM should be based on manufacturer data for the actual equipment being used. The temperature split across the indoor coil can range from 17 to 21 degrees (depending on geographic location) and is calculated by using the sensible heat ratio. The sensible heat ratio is the total load of the structure divided by the sensible load of the structure.

Examples:

	House A	House B
Sensible Heat Gain	32,050 Btu	27,000 Btu
Latent Heat Gain	3,950 Btu	9,000 Btu
Total Heat Gain	**36,000 Btu**	**36,000 Btu**

House A 32,050 Btu/36,000 Btu = **0.89 Sensible Heat Ratio**

House B 27,000 Btu/36,000 Btu = **0.75 Sensible Heat Ratio**

Temperature Difference (TD) Table	
Sensible Heat Ratio	TD
0.75 to 0.79	21°F
0.80 to 0.84	19°F
0.85 to 0.90	17°F

In the examples, House A has less latent load therefore the designed TD across the evaporator is 17°F, requiring a higher CFM than House B.

Use the sensible heat formula:

CFM = Sensible Btu ÷ (TD x 1.08)

According to this formula, House A requires 1,746 CFM and House B requires 1,190 CFM. In comparison, a traditional three-ton system is designed to move 400 CFM per ton, equaling 1200 CFM of indoor air.

The slower the air moves across an evaporator, the greater amount of moisture is removed from the air, lowering the humidity. Depending on the geographic location, the CFM for cooling can range from 360 CFM to 530 CFM per ton. For an installed system to work, the designer matches the sensible and latent loads of the system to the sensible and latent loads of the structure.

When using dual or variable-capacity systems with an **electronically commutated motor (ECM),** the blower motor varies airflow to match the system demand. Excellent humidity control can be achieved with an ECM blower motor and a thermostat that senses temperature and humidity. On these systems, blower motors decrease or increase its speed in order to control the air temperature drop across the evaporator. This allows the evaporator to remove more or less moisture from the indoor air.

Systems using the ECM type blower usually have a circuit board with dip switches or jumpers to select the CFM for cooling and heating. When the fan switch (on the room thermostat for some systems) is set to "on" or "continuous", the ECM blower runs producing half of the cooling CFM until there is a call for cooling or heating.

Checking and Calculating Indoor Air Flow

Many methods can be used for checking airflow delivery from the indoor system. Depending on the needs of the technician, instruments can be purchased for as little as $50 or as much as $3,000.

Basic instruments are used to determine airflow by calculating how fast the air is moving. If the size of an opening and the speed of the air are known, the CFM is easy to calculate.

CFM= Square feet x Feet per minute (Fpm)

Example A: A 12-inch by 12-inch square duct with the air flowing at 500 feet per minute (Fpm) equals 500 CFM.

Example B: A 12-inch by 24-inch square duct with the air flowing at 500 feet per minute (Fpm) equals 1000 CFM.

12 inches **x** 24 inches = 288 square inches

288 ÷ 144 square inches per square foot = 2 square feet

2 square feet **x** 500 Fpm = 1000 CFM

When calculating air flow for a supply or return air grille, use the net free area of the grille not the dimensions of the grille. The net free area of a grille is the actual open space for the air to pass between the louvers. A return air filter grille has an average of 85% net free area.

Motor Speed	External Static Pressure				
	0.1	0.2	0.3	0.4	0.5
High	1,352	1,318	1,260	1,202	1,128
Medium	1,214	1,172	1,123	1,064	1,012
Medium-Low	997	994	960	923	884
Low	757	753	734	704	674

Example:

A 24 inch **x** 24 inch return-air grille has 576 square inches of total area.

576 square inches **x** 0.85 = 490 square inches of net free area

490 square inches ÷ 144 = 3.4 square feet

3.4 square feet would be considered the AK factor of the grille.

Another method is to use the blower performance chart supplied by the manufacturer of the equipment. By using an electronic or inclined manometer, the pressure drop or total static pressure across the indoor air handler can be determined. By plotting the static pressure and blower motor speed tap on the blower performance chart, the approximate CFM can be calculated.

Blower Performance

CFM: Temperature Rise Method with Electric Strip Heat

When using the temperature rise method, run the system until it stabilizes. A heat sequencer can take a few minutes to energize all the elements. When running on emergency heat, set the fan to the "on" position to insure the blower is running at the speed used for normal heat pump operation. The blower speed for emergency heating may be less than that used during normal heat pump operation.

NOTE: This method indicates the approximate CFM for field service. It is not as accurate as using meters designed to measure air flow (such as the air flow hood).

There are five steps for calculating the CFM:

1. Place the fan switch to the on position and the system switch to emergency heat, with the temperature setting high enough to keep the electric heat on.
2. Measure the line voltage and the total amperage draw of the indoor electric heat system to the nearest tenth of a volt and the nearest amp.
3. Measure the average temperature split across the indoor air handler (supply air temperature minus return air temperature).
4. Apply the formula: **Volts x Amperage x 3.41 = Btu**
5. Apply the formula: **Btu** (from step 4) ÷ **1.08** ÷ **ΔT** (TD from step 3) = **CFM**

Example:

230.4 volts **x** 44.8 amps **x** 3.41 Btu/watt = 35198 Btu
35198 Btu ÷1.08 ÷ 25°F ΔT = 1300 CFM

CFM: Temperature Rise Method with Dual-Fuel Heating

The temperature rise method can also be used to calculate the approximate CFM of a fossil fuel furnace. It is important to remember that the input Btu rating should *not* be used; use the net or bonnet capacity in the calculations. The furnace must be set for proper combustion efficiency in order for this method to work. Allow the system to run until it stabilizes. Set the fan to the on position to insure the blower is running at the speed used for normal heat pump operation.

NOTE: This method indicates the approximate CFM for field service. It is not as accurate as using meters designed to measure air flow (such as the air flow hood).

1. Place the fan switch to the on position and the system switch to "emergency heat", with the temperature setting high enough to keep the furnace on.
2. Measure the average temperature split across the furnace (supply air temperature minus return air temperature). Apply the formula: **Net Btu ÷ 1.08 ÷ ΔT = CFM**

Outdoor System Air Flow

The air flow across the outdoor coil varies depending on the manufacturer and model. Some manufacturers build a system with a small footprint moving a lot of air, or a large footprint moving little air. A system with a 650 RPM fan does not make as much noise as a motor running at 1075 RPM.

The CFM for a condensing unit can be acquired from the manufacturer, and can be used to calculate total BTU removed while operating in the cooling mode.

The outdoor design fan speed and CFM are determined by the manufacturer. The fan speed may be controlled by an outdoor thermostat, pressure switch, or electronic speed control. In either case, the CFM must be known to calculate the net BTU capacity while operating in the cooling mode.

Review Questions
Section 3: Airflow

Name: ______________________________ **Date:** ________________

1. To prevent high discharge pressure during the heating cycle, most manufacturers recommend a CFM range of __________to __________ for the indoor unit.

2. The ________ air moves across the indoor coil, the more moisture will be removed.

3. Excellent humidity control can be achieved with a/an __________ and a/an __________ that senses temperature and humidity.

4. Use the following information to calculate the CFM: 12" x 12" square duct with air flowing at 500 FPM.

5. What is the following equation, "Volts x Amperage x 3.41" used to calculate?

6. Use the temperature rise method with strip heat to determine the __________.

7. The final CFM for Cooling should be based on what data?

8. The temperature split across the indoor coil can range from _____F° to _____°F for a heat pump in the cooling cycle (depending on the geographic location).

9. What is the formula Sensible heat Btu divided by (1.08 × the temperature difference) used to calculate?

10. When calculating the air flow of a supply or return an air grille, which should be used; gross area or net free area?

11. What are the necessary steps to follow when using the Temperature Rise Method with an Electric Heat Strip unit to find the CFM for cooling?

Lab Assignment 1

Section 3: Airflow

Name: ____ **Date:** ____________________

Objective: Verify blower operating conditions and air flow quantity through an air handler based on static pressure. The system must have a return air with a filter/grille and a supply plenum.

Directions: Use a manometer to measure total system static pressure and compare measurement to manufacturer's blower performance chart.

NOTE: Perform the following task using all safety procedures!

System

Brand: _________________________

Model Number: ______________________ Capacity: ________________ Btu

1. Measure return air plenum static pressure and record. _______________ "W.c.

2. Measure supply air plenum static pressure and record. _______________ "W.c.

3. Record total static pressure. _______________ "W.c.

4. Record the blower speed tap or selection. (High-Low etc.) ____________________

5. Does measured static pressure match a pressure listed on the manufacturer's chart or does it fall between listed pressures? □ Listed □ In-between

6. CFM based on static pressure measurement _______________CFM

NOTE: Show work or math used for calculating CFM when the pressure is between manufacturer's listed pressures.

Lab Assignment 2

Section 3: Airflow

Name: ______________________________ **Date:** ____________________

Objective: Verify blower air flow quantity through an electric heat air handler using the temperature rise method.

Directions: Use electrical and temperature measurements to calculate system CFM.

NOTE: Perform the following task using all safety procedures!

System:

Brand: ______________________ Model Number: ______________________

Capacity: ________________ Kw Calculated CFM: ______________________

Measurements:

Voltage: ________________ Amperage: ________________

Supply Air Temperature: ______________ Return Air Temperature: ______________

Show math for calculations below:

Objectives

Upon completion of this section, the participant will be able to:

1. name the four defrost methods
2. describe the differences between the 4 defrost methods

Temperature Differential

On small PTACs and window units, a dual-sensing thermostat is used to sense the temperature differential between the outdoor coil and outdoor air. As frost builds up on the outdoor coil, the temperature differential decreases and defrost mode is initiated.

Temperature Differential Thermostat

Time and Temperature

Time and temperature is the most common defrost method. Time intervals for starting the defrost cycle can be set for 30, 60 or 90 minutes by moving a jumper wire. A set of test terminals can be jumped out to start defrost when servicing the system. Time and temperature must be satisfied to initiate defrost. Either the timer or the thermostat can end the defrost cycle. To prevent the compressor from over-heating, most electronic timers shut off the defrost cycle after 10 minutes if the thermostat fails to open.

Defrost thermostats usually close below 30°F and open between 50°F and 70°F, depending upon the system design.

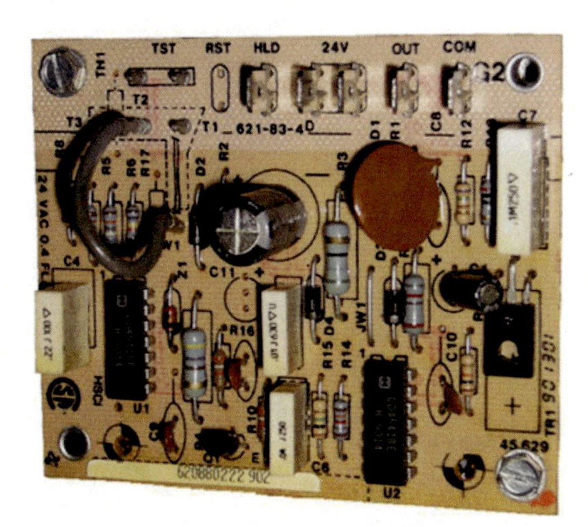

Electronic Defrost Control

Temperature and Pressure

This is a combination control made with a diaphram activated pressure switch and a defrost thermostat. The pressure switch senses either a positive or negative pressure change across the outdoor coil as it ices up. The defrost thermostat sensing bulb monitors the temperature of the outdoor coil. When both the pressure switch and the defrost thermostat close, the defrost cycle is started. Ending the defrost cycle is handled by the defrost thermostat only.

Temperature/Pressure Defrost Control

Time, Temperature and Pressure

Time, temperature and pressure method uses a combination of all three controls. To initiate the defrost cycle, all three - time, pressure across the coil, and coil temperature - must be satisfied. Either the coil defrost thermostat or the timing control can end the defrost cycle. This method is called a demand defrost. By using all three methods the defrost cycle is delayed until efficiency is compromised.

Time/Temperature/Pressure Defrost Control

Review Questions
Section 4: Defrost Methods

Name: ________________________________ **Date:** __________________

1. Which defrost control is normally used on small window units and PTACs?

2. The defrost control for a residential air source heat pump uses _________ and ____________ to initiate defrost.

3. In the temperature and pressure defrost method, the defrost cycle is ended by the __________.

4. Which defrost method is commonly referred to as demand defrost?

5. What is used to defrost the outdoor coil for a heat pump?

6. Write a brief sequence of "defrosting" a heat pump by the time and temperature.

7. Approximately what temperatures does a defrost thermostat close and open its contacts?

8. What is necessary for the technician to do in order to make a solid state defrost board go into defrost during testing?

Lab Assignment 1
Section 4: Defrost Methods

Name: ______________________________ **Date:** ________________

Objective: Familiarization and knowledge of troubleshooting defrost controls.

Directions: Identify heat pump defrost control type, and general features.

Defrost Type, check one: □ Mechanical □ Electronic

Available cycle times, check all that apply: □ 30 minutes □ 60 minutes □ 90 minutes

Defrost control temperature sensor, check one: □ Thermostat □ Thermistor

Answer the following that applies to system :

Defrost termination: Thermostat temperature range open and close: _______°F _________°F

Defrost thermistor resistance: ________________Ω at _____________°F Temperature

Does the defrost control have a test mode: □ Yes □ No

NOTE: Perform the following task using all electrical safety procedures!

1. With power to the equipment off, disconnect the outdoor fan motor power leads from the defrost board or defrost relay and tape ends to prevent a short circuit.
2. Connect and insulate an electronic thermometer sensor as close as possible to the defrost thermostat or thermistor.
3. Apply power to the system, operate in the heating mode.
4. Record the temperature at which the defrost thermostat closes: _______________°F.
5. Jumper test terminals on the defrost board and

   record the temperature as the defrost terminates: _______________°F.
6. Turn off system and all power.
7. Remove test equipment and reconnect all electrical connections removed for the test.

Objectives

Upon completion of this section, the participant will be able to:

1. define coefficient of performance (COP)
2. calculate balance point
3. calculating balance point in dual fuel systems
4. calculate multiple outdoor thermostat settings

Coefficient of Performance

Coefficient of Performance (COP) is the efficiency of an air-to-air heat pump expressed as a ratio of heat energy transferred to units of energy used. The COP of resistance heat is 1:1 and 100% efficient. Heat pumps are capable of producing a COP range as high as 3:1 to 3.5:1 or as much as 350% efficient. Water-to-air heat pumps can produce a COP of 4:1. This high COP makes heat pumps much more economical than an electric resistance heat source.

The COP rating for air-to-air heat pumps is based upon two outdoor operating temperatures: 47˚F and 17˚F. The Coefficient of Performance decreases as the outdoor temperature decreases. In fact, the COP can decrease to the same level as resistance heat, 1:1. This happens when the outdoor ambient temperature is between 10˚F to 0˚F, depending on make and model. When the COP reaches 1:1 some manufacturers program the heat pump system to cycle off leaving the entire heating task to the auxiliary resistive heaters and indoor blower motor. Other manufacturers choose to continue running the entire system, even at 0˚F, to prevent refrigerant migration to the compressor.

Balance Point Calculations

The Balance Point is the lowest outdoor temperature at which a heat pump can satisfy the heating load of the structure. The balance point is not the temperature at which the COP is 1:1 Heat pumps are sized to the cooling load in most applications, not the heating load. For this reason, the heat pump may not satisfy the entire heating load of a structure, and balance point must be considered. When the outdoor temperature drops below the calculated balance point, some form of supplemental heat becomes necessary to maintain a comfortable indoor environment. A heat pump operates continuously when the outdoor temperature is below the calculated balance point.

Resistance heaters installed as supplemental heat, are staged to come on last and go off first to minimize operating costs. Most systems employ resistance heat as an emergency backup heating system in case the heat pump fails. It is recommended to have enough supplemental heat to equal 80% of the designed heat load.

Dual Fuel Systems

Dual Fuel systems require the heat pump to cycle off before energizing gas or oil heaters at the balance point. The heat pump coil is located in the air stream after the fossil fuel heater. Cycling the furnace and heat pump on at the same time would cause high discharge pressures and temperatures.

In the summer, when the heat pump is operating in cooling mode, air leaving the evaporator coil is approximately 100% relative humidity. The evaporator coil is located after the fossil fuel heat exchanger to prevent the heat exchanger from rusting.

Balance Point Calculation Example

Calculate the outdoor thermostat temperature setting for each installation. The load of the structure, geographic location, and the model of the system are used to determine the balance point setting.

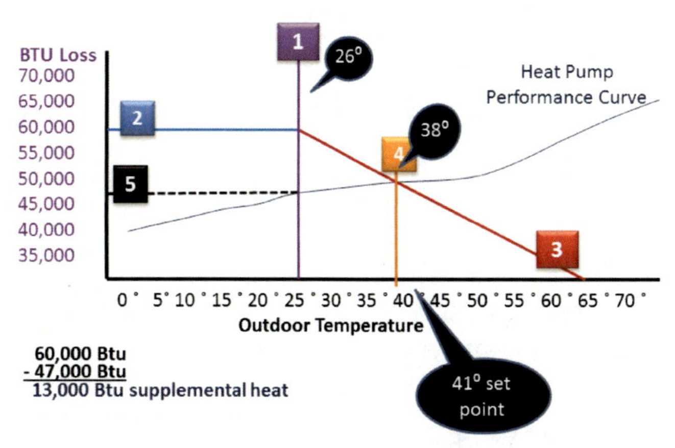

Example:

Line 1. The outdoor design temperature for the geographic location.

Line 2. The heat loss of the structure at the designed temperature for the area. (*Note*: This can only be determined by a heat loss study, such as "Manual J".)

Line 3. The line indicating heat loss, as outdoor temperature increases. 65°F is considered the temperature at which there is no need for heating or cooling. The line is drawn from 65°F, up to the intersection of the design temperature and heat loss of the structure. Drawing a vertical line from an outdoor temperature to this line, then to the left indicates the heat loss for the structure at that outdoor temperature.

For example, at 50°F outdoor temperature, the structure's heat loss would be approximately 41,000 Btu.

Line 4. This line is the balance point line. 39˚F is when heat pump capacity is equal to the heat loss of the structure.

Line 5. The intersection of the heat pump capacity and the outdoor design temperature. This line indicates the capacity of the heat pump at that designed temperature. The left side of this line indicates the Btu capacity of the heat pump. Subtract it from the structure heat loss. In this example, the 60,000 BTU structure heat loss minus 47,000 BTU output of the heat pump equals 13,000 BTU.

13,000 BTU is the supplemental heat required to satisfy heat loss for at design temperature. In this example, 39˚F is the actual balance point. The outdoor thermostat should be set three degrees higher or at 41˚. This allows the supplemental heat to be energized, providing enough additional heat to allow the system to cycle off.

To calculate the kilowatts of heat needed, use the following formula:

Btu ÷ watts = Kw

So: 13,000 Btu ÷ 3,413 watts = 3.8 Kw

Outdoor Thermostats

The outdoor thermostat (ODT) senses outdoor ambient air temperature and is set to close its contacts three degrees above calculated balance point. When the ODT senses temperature below the calculated balance point, the contacts close allowing the heat strips to be energized if the room thermostat 2nd stage is calling for heat. This provides supplemental heat in addition to the heat pump, and prevents the unit from running continuously during cold ambient conditions. Eventually, as the outdoor temperature rises above the ODT set point, the ODT contacts open, returning the system to heat pump operation only.

Outdoor thermostats can be used to energize multiple stages of heat. Setting multiple ODTs for lower temperatures can keep the strip heat off until predetermined temperatures are reached.

Outdoor thermostats are valuable to keep strip heaters off when not necessary thus helping to maintain a higher efficiency. If the occupant turns the room thermostat up more than two degrees higher than actual room temperature,

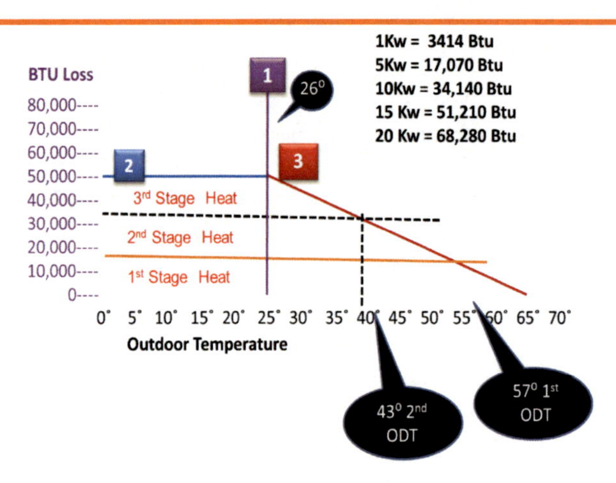

Fifteen Kilowatt Strip Heat Calculation

the 2nd stage bulb tries to bring on all the electric strip heaters if the system does not have an ODT.

Strip heaters can be controlled by an ODT for each stage or bank of heaters. If the outdoor temperature is above the ODT setting, the room thermostats' 2nd stage bulb cannot bring all of the strip heaters on at one time. Just the strip heaters the ODT allows can operate as shown below.

Calculating Multiple Outdoor Thermostat Settings
Refer to the diagram above.

Line 1. A vertical line drawn up from the outdoor design temperature for the geographic location in which the heat loss calculation was performed.

Line 2. Horizontal line indicating the Btu heat loss of the structure at the designed temperature for the area. The line is drawn from the left Btu loss to the right intersecting line 1. (Note: This can only be determined by a heat loss study, such as "Manual J".)

Line 3. Is drawn from the 65°F point on the outdoor temperature line on an upward angle to the intersection point of line 1 and 2. This line indicates the increase in heat loss of the structure as the outdoor temperature drops.

Line 4. Horizontal line from the Btuh quantity of the 1st stage electric heat intersecting line 3. The vertical line at the point of line 3 and 4 (54°F) is the lowest temperature the 1st stage (In the example 5Kw) will heat the structure.

The outdoor thermostat would be set at 56°F to 57°F to energize the next stage of heat.

Line 5. Horizontal line from the Btuh quantity of the 1st and 2nd stages of electric heat intersecting line 3.

The vertical line at the point of line 3 and 5 (40°F) is the lowest temperature the 1st and 2nd stage (In the example 10Kw) will heat the structure.

The outdoor thermostat would be set at 42°F to 43°F to energize the next stage of heat.

In this example 15Kw of heat satisfies the heat loss of the structure (50,000 Btuh)

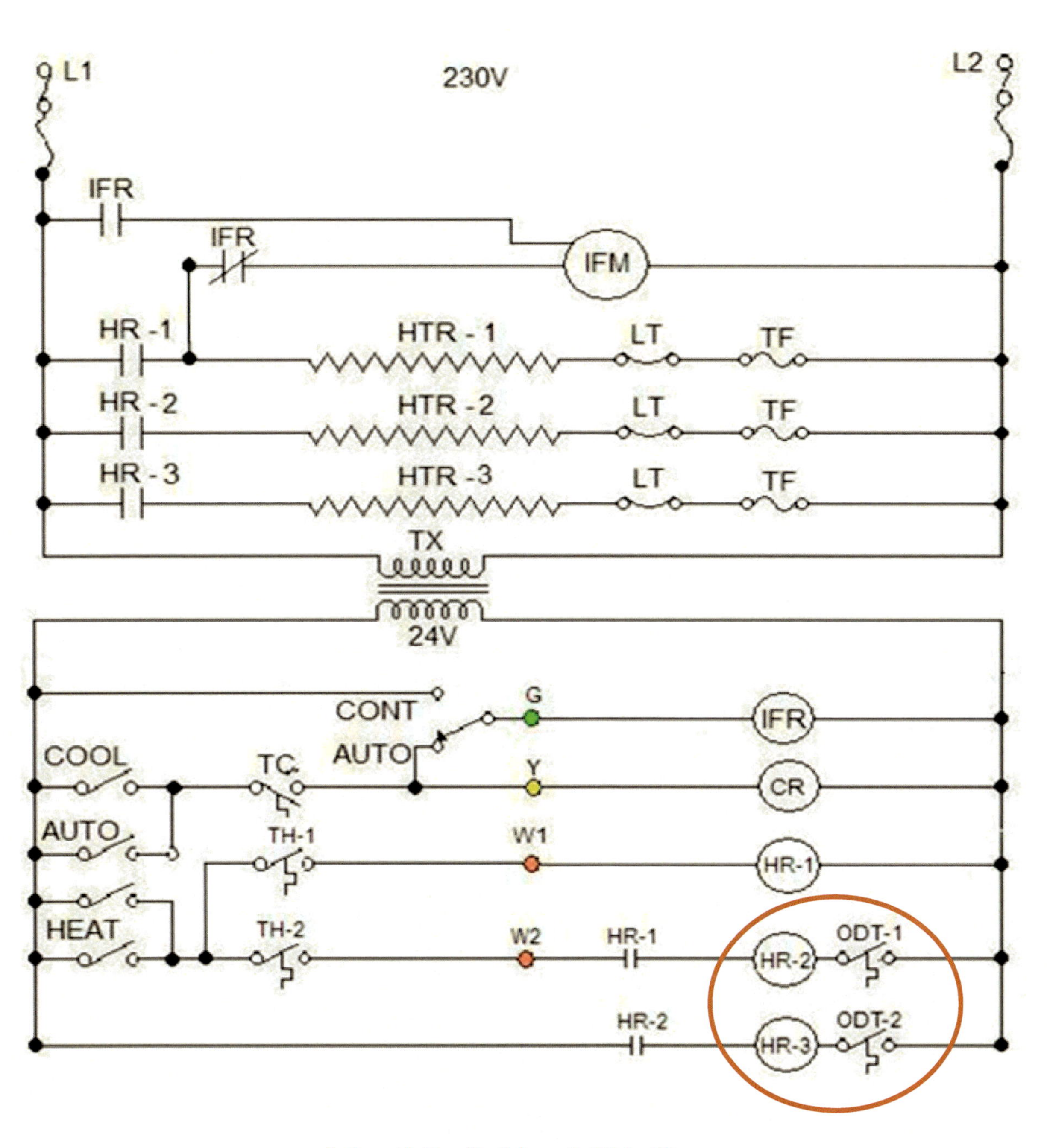

Indoor Air Handler Schematic Wiring Diagram

Review Questions
Section 5: Balance Point

Name: ______________________________ **Date:** __________________

1. COP is the abbreviation for __________.

2. The COP rating for an air-to-air heat pump is based upon two outdoor operating temperatures: __________ °F and __________ °F.

3. As outdoor temperature decreases, the COP __________.

4. The lowest outdoor temperature at which a heat pump can satisfactorily heat the structure is known as the ____________.

5. In a dual fuel system, balance point temperatures should be calculated to allow the heat pump to cycle __________ before the gas or oil heater cycles __________.

6. The outdoor thermostat (ODT) should be set to close its contacts at __________ degrees above the calculated balance point.

7. On most residential air source heat pumps, the defrost cycle turns on what type of heating assistance?

8. What is the outdoor temperature when heating or cooling should be not needed?

9. To change BTUs to Kilowatts, you use what formula?

10. What happens when only the indoor thermostat controls the secondary heater for a heat pump?

Section 6: Secondary Heat

Objectives

Upon completion of this section, the participant will be able to:

1. name the types and purpose of secondary heat
2. identify the purpose and operating differences of emergency heat

Secondary heat can be electric, oil or gas. The most commonly used is electric. During the system design process designers must calculate the additional BTUh of heat needed by a heat pump when operating at low outdoor temperatures. If the auxiliary heat is electric, the calculated BTUh can be changed to watts or Kilowatts (KW). The electric heater(s) provide supplemental heat as needed..

When a fossil fuel furnace is used as the secondary heat source, the heat pump must be turned off.

Auxiliary Heat

Auxiliary heat should be approximately eighty percent of the total load, calculated at the design conditions. Auxiliary heaters can be electric resistance, gas, oil, or solar.

Supplemental Heat

Supplemental is the amount of resistance heat needed to handle the design load after the balance point is reached.

Electric Resistance Heat

Electric resistance heat provides heat to a space during the defrost cycle, delivers supplemental heat when there is a demand for second stage heating, and supplies emergency heat if the heat pump fails to operate properly. It may also be used to meet the total heating requirements of the space.

Example: If a heating load calculation indicates a deficiency of 16,380 Btuh, then approximately 4.8 KW of resistance heat must be added.

$$\text{KW required} = \text{Btu heat loss} \div 3413$$

Resistance heaters are available in various sizes and are designed to be energized in stages. Sequencers are used to energize the stages.

Auxiliary heat and supplemental heat describe how electric heat is used with the heat pump system. Some manufacturers turn the heat pump outdoor unit off when the heat pump is producing a COP of 1 or a 1:1 ratio. The auxiliary electric resistance heat must provide heat for the structure. As shown previously, this resistance heating system must provide enough heat for the structure or 80% of the total heat loss of the structure.

Supplemental heat comes from the same heat source as auxiliary heat, but is called supplemental because only part of the heaters are energized or staged on to supplement the heat pump when the balance point is reached. On demand for second stage heat, the supplemental heaters are energized if the outdoor thermostat(s) confirms the temperature requirement. Depending on the number of supplemental heaters required or available, systems may have one or more outdoor thermostats.

When the heat pump is in defrost mode, the electric resistance heater(s) are turned on to prevent cold air from the indoor coil being delivered to the conditioned space. Supplemental heaters prevent the occupant from feeling a blast of cold air anytime the heat pump cycles to defrost mode.

Emergency heat is the same electric resistance heating system, but operates somewhat differently. When the system is switched to emergency heat mode, the outdoor unit cannot operate. Instead, the first stage of electric heat comes on, and if the indoor temperature is not satisfied or continues to decrease , the second stage of electric heat is activated by the thermostat's second stage. The outdoor thermostat is bypassed when the systems thermostat is switched to emergency heat.

When the manufacturer designs the system to turn the outdoor unit off at a very low outdoor temperature, the electric resistance heat has to satisfy the total heating requirement of the space.

Review Questions
Section 6: Secondary Heat

Name: ______________________________ **Date:** ________________

1. What are three common types of secondary heat?

2. Auxiliary heat should be approximately __________% of the structural load, calculated at the design conditions.

3. What does placing the thermostat to Emergency Heat do to the outdoor section of a split system?

4. How should electric secondary heater(s) come on?

5. What are the functions in which the secondary heaters operate?

Objectives

Upon completion of this section, the participant will be able to:

1. identify the indoor system components
2. identify outdoor system components
3. identify components and the sequence of operation from a heat pump ladder diagram

Indoor Components

The main control for the system is the indoor thermostat. As shown in previous sections, the thermostat can be either mercury bulb or solid state digital type. The thermostat is normally wired with color coded wiring, and operates on low voltage (24VAC). The wiring size should be 18-gauge, unless the local codes prevail. It should be mounted 52" above the floor, near the main return and on an interior wall (not on an outside wall or by the structure's entrance/exit). The thermostat controls the room temperature, the system mode of operation: heat, cool, or emergency heat, and the indoor blower. A switch controls the blower fan motor. When in the ON position, the blower runs continuously and in the AUTO position, the fan cycles with demand for cooling or heating.

Another important feature of this thermostat is the "Emergency Heat Mode". When the system switch is moved to the Emergency Heat position, the thermostat does not make a circuit to the contactor bringing on the compressor and outdoor fan motor. Instead, the thermostat brings on the 2nd stage of heat to keep the structure warm. The indoor fan motor may run at a slower speed while the system in Emergency Heat mode, unless the fan switch is ON.

Most indoor thermostats have a subbase mounted on the wall before the control-section is attached. The subbase has terminals for the low-voltage control wiring. The color-coded low voltage wires are connected to the proper terminals on the subbase. It is very important the opening around the low voltage cable entrance is sealed with a filler material. Cold or warm air can pass through the wall opening and cause the thermostat to function at the wrong temperatures.

These color codes and terminal identifications are used by many heat pump and thermostat manufacturers:
Red to R (24V feed),
Green to G (Fan)
Orange to O (Reversing valve)
White to W1 (Heat, 2nd)
Brown to W2 (Heat, 3rd) optional
Yellow to Y (Contactor)
Blue to C (24V Common)
Black to E or X2 (Emergency heat for some heat pumps)

Transformers reduce line voltage to 24 volts for control operation. Low voltage controls are easier to wire and easier to manufacture.

Transformers are rated in Volts X Amperes (VA). Some older systems used two 20VA transformers, one in the indoor unit and one in the outdoor unit. Most equipment manufactured today have only one 40 to 50VA transformer in the indoor unit which provides low voltage for both indoor and outdoor sections. This simplifies the wiring, making it easier to troubleshoot and service.

The indoor blower normally has a multi-speed PSC motor. The motor runs at speeds selected during installation.

Electronically Commutated Motors (ECMs) are often used in high efficiency indoor systems instead of PSC. This motor can vary speed to match airflow requirements of the structure or mode of operation. This motor has been explained in earlier sections.

Sub-Base

Indoor Thermostat Sub-Base

Indoor Thermostat

The blower motor relay is located in the air handler or indoor section of the system. It is the primary control for the blower motor. Coil voltage is 24VAC and is energized by the G terminal on the indoor thermostat. To improve efficiency, some systems utilize a time-delay fan relay which provides approximately a 30 second delay on and off.

Sequencers are used to bring on stages of electric strip heat and may control the blower. Sequencers operate by 24VAC or high voltage control elements. Sequencers are time-delay switches that open or close at various intervals to bring on or turn off stages of electric heat. The resistance heating elements must not come on without the indoor blower motor running. Without the proper air moving over them, the elements become bright red and overheat. The stages of electric heat have two safeties which open when the heater ambient temperature is excessive.

The stages of electric heat have two safeties that open when the heater's ambient temperature is too high. These are the thermal-cutout and fusible-link safety controls. The thermal-cutout is a bi-metal control that opens its contacts at a preset temperature. When the contact opens, it disconnects the high voltage from the heating element and resets automatically when the temperature falls below a preset temperature. Should the temperature rise to or above the temperature at which the fusible links are rated, the link will open and not re-set.

Outdoor Components

The outdoor control wiring can be low voltage, high voltage or both.

The contactor in the heat pump controls the compressor and outdoor fan. This component is energized by the thermostat in both heating and cooling modes. Most contactors in residential units are controlled by low voltage (24VAC). On larger systems, the thermostat uses a 24VAC pilot relay to control line voltage power to the contactor coil. The contactor may be a single or a double-pole for single-phase units, and double or three-pole for three-phase units.

The defrost control cycles the system into defrost mode as needed. The defrost control on most residential heat pump systems is the solid state type, which can control high and low voltage loads.

The defrost relay (when used) either energizes or de-energizes the reversing valve. During defrost, it de-energizes the outdoor fan and energizes the supplemental heat.

The reversing valve relay (when used) is used when the reversing-valve solenoid coil voltage is higher than twenty-four volts, or the defrost control cannot directly control the line voltage or current that is required for valve operation.

Electrical Control Diagrams

Shown at right is a ladder diagram of a generic heat pump. In this example, the heat pump is a single-phase, 230-VAC pump with a single-speed, single-capacity compressor motor (PSC), and a single set of contactor contacts (CR) controlled by the CR contactor coil located in the low-voltage section at the bottom right of the diagram. This heat pump also has a low-voltage indoor thermostat, which has one cooling stage (TC) and heating stages (TH-1 and TH-2). The system has three electric heating elements: HTR-1, -2, and -3, which are controlled by sequencers HR-1 and HR-2 contacts. Note the safeties LT (thermal cutouts) and TF (fusible links) for each heating element.

The crankcase heater is on continuously, as shown in this heat pump illustration. The defrost control shown is a solid-state type with only low voltage running to and from it. This system also has an ODT (outdoor thermostat) in series with TH-2, which keeps all the resistance electric heat off unless the outdoor ambient temperature setting of the thermostat is reached. The defrost control or the emergency heat relay will bypass the outdoor thermostat and energize the heaters as needed.

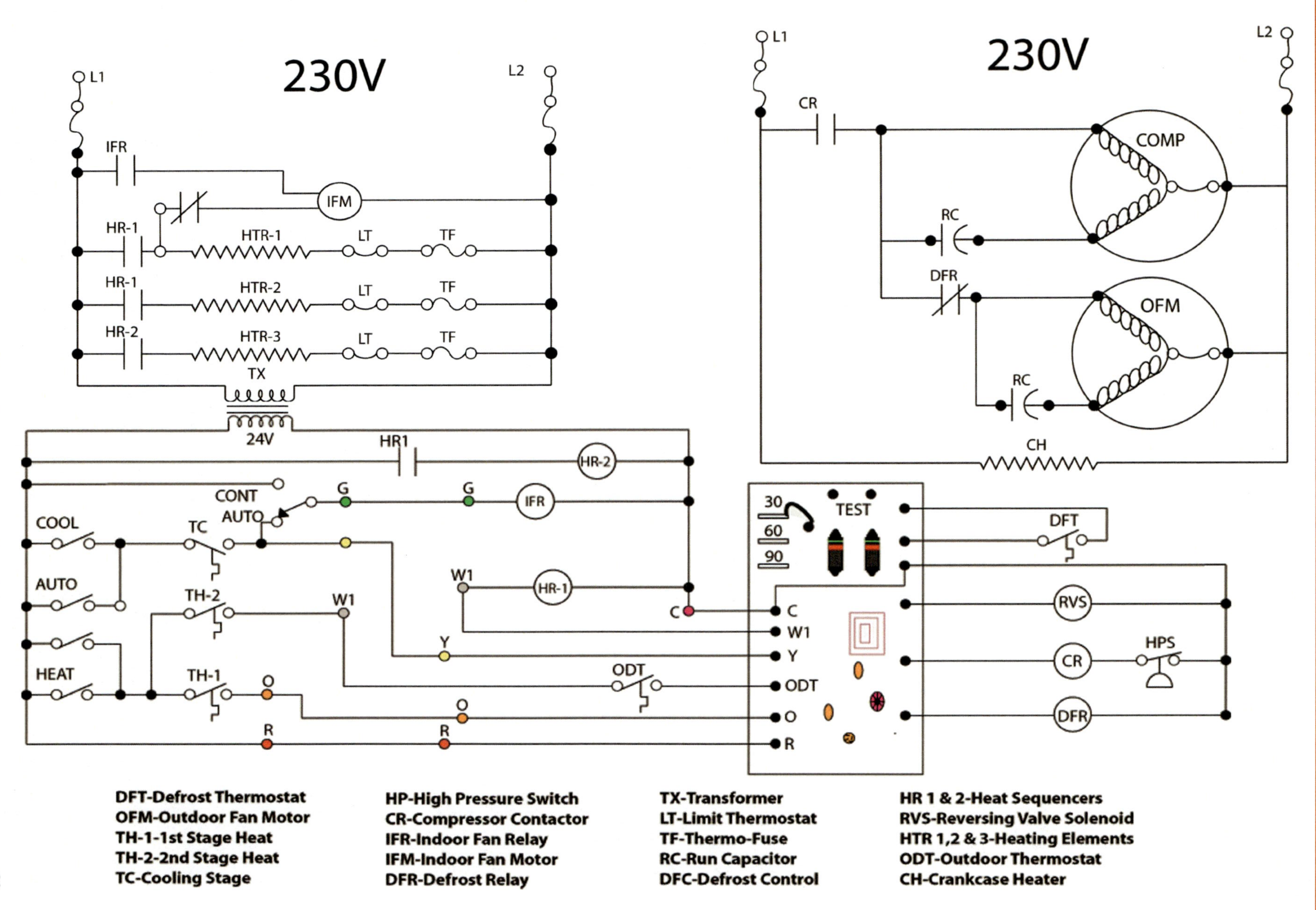
230V
L1
L2
IFR
IFM
HR-1
HTR-1
LT
TF
HR-1
HTR-2
HR-2
HTR-3
TX
24V
HR1
HR-2
CONT
AUTO
G
IFR
COOL
TC
AUTO
TH-2
W1
HR-1
HEAT
TH-1
O
R
Y
C
ODT
230V
CR
COMP
RC
DFR
OFM
CH
30
60
90
TEST
DFT
RVS
HPS
DFR
DFT-Defrost Thermostat
OFM-Outdoor Fan Motor
TH-1-1st Stage Heat
TH-2-2nd Stage Heat
TC-Cooling Stage
HP-High Pressure Switch
CR-Compressor Contactor
IFR-Indoor Fan Relay
IFM-Indoor Fan Motor
DFR-Defrost Relay
TX-Transformer
LT-Limit Thermostat
TF-Thermo-Fuse
RC-Run Capacitor
DFC-Defrost Control
HR 1 & 2-Heat Sequencers
RVS-Reversing Valve Solenoid
HTR 1,2 & 3-Heating Elements
ODT-Outdoor Thermostat
CH-Crankcase Heater

Review Questions
Section 7: Electrical Control Wiring

Name: ________________________________ **Date:** ____________________

1. The main control for a heat pump system is the __________.

2. The control wiring size for the indoor thermostat should be a minimum of __________ gauge.

3. What is the rating for most control transformers used in a residential heat pump?

4. In emergency heat mode, the fan speed will usually __________.

5. The control wiring for the thermostat is normally connected to the thermostats' ________.

6. The type of diagram that shows the components between the source lines is called a/an __________.

7. Many manufacturers use the same designations for the 24 volt colored wiring; the following wiring is for a seven wire connection. Match the wiring to the correct components.

Red	Indoor fan
Green	Contactor for Compressor and outdoor fan
Orange	24 volt hot terminal from transformer
Blue	Common terminal from transformer
Yellow	Secondary heaters
Black	Reversing Valve
White	Emergency heat

8. What is a Sequencer and when is it used on a heat pump system?

9. Approximately how many electric heat strips are energized with one set of contacts on a sequencer?

10. What are the safeties on electric heaters that prevent overheating and which safety must be replaced after it functions?

Objectives

Upon completion of this section, the participant will be able to:

1. determine proper refrigerant pipe size
2. understand why liquid lines must be properly sized
3. identify reasons and methods to insulate refrigerant line piping

Refrigerant Pipe Sizing

Refrigerant pipe size is determined by length, horizontal/vertical runs, type of refrigerant used, and system capacity. The vapor line is normally sized to provide a minimum velocity of 700 feet per minute (fpm) in a horizontal run and 1,500 fpm in a vertical run in order for the oil to return to the compressor properly. The pressure drop allowed for the vapor line (from indoor coil to outdoor unit) is not the same for all refrigerants. The pressure drop in the vapor line should equal less than a 2°F change in temperature.

Example: A **refrigerant 22** evaporator with an operating temperature of 45°F has a suction pressure of 74.5 psig.

A 2°F change or drop in temperature equals 43°F and 73 psig from the pressure/temperature chart.

The difference in the two pressures equals a 1.5 psi maximum pressure drop.

Using **refrigerant 410a** and the same temperature of 45°F (less 2°F = 43°F) the pressures would be 130.7 psig and 125.9 psig. This would be a maximum pressure drop of 4.8 psig. for the 410a system.

As demonstrated above, type of refrigerant, type of oil used, and evaporator operating temperatures affect the oil return process.

Depending on the system design, the refrigerant line pressure drop may be increased to maintain the velocity to assure proper oil return to the compressor. Different sizes of tubing may be connected together to maintain velocity and low pressure drop. Remember, an *increase* in pressure drop lowers the system's capacity, and a low refrigerant velocity prevents the oil from returning to the compressor, causing compressor failure. Always use the manufacturer's recommendations when sizing refrigerant lines.

Do not use refrigerant line sets that are longer than necessary. Extra-long lines, even laid horizontally, can become an oil trap and cause problems with entrainment of oil back to the compressor. Long lines also increase the total refrigerant charge, which can cause compressor problems such as flood back or refrigerant migration to the compressor.

The chart shows the unit size in BTUh, vapor line diameter and equivalent cooling capacity losses due to line length in feet (FT). The 60,000 BTUh unit, using a 50 FT vapor line diameter of 7/8", has approximately a 7% cooling capacity loss. The same unit with a 1-1/8 inch vapor line has only a 1% cooling loss. The smaller diameter (7/8") vapor line creates more velocity for the oil entrainment process but has greater cooling loss.

In the chart shown below, the equivalent line length (FT) is the estimated length and includes pressure drop for fittings and elbows.

Estimated Percentage of Cooling Capacity Loss					
BTUH	Vapor Line Diameter	Equivalent Line Length			
		50 ft	75 ft	100 ft	125 ft
24,000	5/8	6	9	13	16
	3/4	0	1	1	2
30,000	5/8	6	8	10	13
	3/4	2	3	4	5
36,000	3/4	7	10	14	17
	7/8	2	4	6	8
42,000	3/4	7	10	13	17
	7/8	3	4	6	7
	1-1/8	0	0	1	1
48,000	3/4	10	14	18	22
	7/8	4	5	7	9
	1-1/8	0	0	1	1
60,000	7/8	7	9	11	14
	1-1/8	1	2	2	3

Liquid Lines

Liquid lines are affected by size and length. If a liquid line is too small, pressure drop is excessive and restricts refrigerant flow. Liquid inside the line may start "change of state" and flash to vapor before reaching the metering device. This causes many problems including excessive superheat and very low cooling capacity. On the other hand, a liquid line that is too large results in charging problems (overcharged or undercharged) because it can hold refrigerant and possibly starve or rob the evaporator. Liquid line velocity is normally less than 300 FPM. Follow manufacturer specifications for refrigerant lines.

Section 8: Refrgerant Piping

Pipe sizing is one of the most important parts of an installation in today's high-efficiency equipment. With the advancement of dual capacity or variable speed compressors, velocity changes dramatically. The system may need to be designed with a higher pressure drop to ensure oil return at lower capacities. It may be necessary to use double risers and traps in the suction line for installations with a long vertical rise.

Refrigerant Pipe Installation and Insulation

There are multiple reasons to insulate the refrigerant vapor line. During the cooling cycle, the vapor line must be insulated to prevent condensation and increase in super heat. During heating cycles, insulation prevents a system capacity loss.

Always insulate the vapor line from the outdoor unit to the indoor coil with 1/2" to 3/4" thick closed cell foam insulation. Local codes may dictate the minimum thickness or "R" value for vapor line insulation. Beyond code requirements, insulation needs are determined by temperatures and conditions around the refrigerant lines. When the air handler is installed in an attic, capacity is improved by insulating the section of liquid line in the attic. The suction line and liquid line can gain superheat if exposed to attic high temperatures.

Secure the line set according to local codes to prevent vibrations, and make sure the horizontal runs are sloped toward the compressor to enhance the oil flow. An easy way to remember the function of the lines while in the heating or cooling mode:

The SMALL LINE is always the LIQUID LINE and is always at HIGH pressure.

The LARGE LINE is always the VAPOR LINE, but is NOT always at low pressure.

CAUTION:
The large line contains low-pressure vapor during the cooling and defrost cycles and high-pressure vapor during the heating cycle.

Many technicians learned this the hard way. Connecting the compound gauge (Blue) to the large vapor line in cooling mode, when the system is changed to heating mode, the large vapor line instantly fills with high pressure vapor. The sudden surge of high pressure usually damages the compound gauge and it has to be replaced. Manufacturers usually have a schrader valve connected to compressor suction for service access.

Review Questions
Section 8: Refrigerant Piping

Name: ______________________________ **Date:** ________________

1. The pressure drop in the vapor line should equal less than a __________ °F change in temperature.

2. Liquid lines are affected by __________ and __________.

3. What may occur if the liquid line is too small?

4. The small line is always the __________ line and is always at __________ pressure.

5. The large line is always the __________ line, but is NOT always at low pressure.

6. Approximately what size liquid and suction line would be used on a two-ton heat pump with a 20-foot line set?

7. Approximately what size liquid and suction line would be used on a 4-ton heat pump with a 40-foot line set?

8. What potential problems may occur if the refrigerant lines are too small?

9. How much refrigerant velocity should the vapor line have in a horizontal run?

10. What is the recommended velocity for the liquid line?

Objectives

Upon completion of this section, the participant will be able to:

1. identify basic indoor and outdoor installation requirements

Indoor Systems - Key Points

The indoor section of a split system heat pump can be located in different places depending on the structure and suitable areas available. Air handlers may be in an attic, crawl space, indoor closet, garage or basement. They may be mounted horizontally or vertically.

Some basic installation requirements include:

- Install the air handler in a central location providing required clearances for operation and service.
- The air handler should be placed in a location that will keep the duct runs as short as possible to insure proper airflow and reduce costs.
- The strength of structural members must be adequate to hold equipment.
- Seal all return and supply air duct connections to prevent air leakage in or out of the system.
- Level the system to allow for proper condensate drainage.
- Install an auxiliary drain pan under the air handler if mounted in an attic to prevent water damage to the ceiling in case of a blocked drain. The auxiliary drain pan must have drain piping to outdoors.
- The air handler condensate drain piping must have a trap if located on the negative pressure side of the indoor coil. The drain line must drop two to three inches before entering the trap on a residential unit, and four inches on a commercial unit.
- Follow the manufacturer's installation instructions and all local codes.

Installations and Special Instructions

Horizontal air handlers are commonly used in attic crawl spaces when houses are built on a concrete slab. Locate the air handler to provide easy access to the indoor blower, auxiliary heaters and wiring. Install an auxiliary drain pan with a drainpipe to outdoors. Mount the air handler on proper mounting pads to absorb noise and vibration.

Horizontal air handlers can be mounted in a crawl space or basement by attaching to the floor joists using threaded piping and angle iron. For safety, most codes require a clear service access of 30" in front of the unit. If the basement floor does not have a drain, install a Condensate Pump to remove the condensate.

Remember:

- In humid summer months, gallons of condensate are extracted from the air inside the structure. Condensate can be routed to a dry well outside the structure. A dry well consists of a hole that has been dug in the ground, with sand and gravel inside.
- Vertically mounted air handlers are very versatile; follow the previous instructions given for drains and drain piping.
- *Service access* must always be considered, regardless of location, and all vibration and fan motor noises can be prevented by installing isolation pads.
- With increased use of HFC refrigerants and ester-based oils, it is more important than ever to use nitrogen when brazing refrigerant lines to prevent contamination. Nitrogen prevents oxidation when brazing and helps purge moisture from the system.
- Nitrogen and a trace amount of the appropriate refrigerant should be used to leak test the tubing connections.

SAFETY NOTE:
When HFCs or CFCs are exposed to an open flame, phosgene gas is a bi-product. The technician must avoid breathing or burning these extremely toxic fumes. The technician's exposure to this toxin can be minimized by using nitrogen during the brazing process.

Outdoor Systems - Key Points

Provide proper clearances from the structure.
Install above the normal snow line.
Install level to provide for defrost water runoff.
Install the unit away from prevailing winds.

Before the outdoor unit is installed, consider these important factors:

- **Wiring:** Where the power supply will be located, what size wiring will be needed, and the fuse/breaker size. Follow NEC guidelines for proper wiring sizes and requirements.
- **Noise:** The unit should be located so that noise will not be a problem, preferably at the side or back of the structure, although in some instances the front of the structure may be used. The unit may also be located on the rooftop in structures like apartment complexes.
- Ductwork: If the system is a package unit, the ductwork must connect to the unit. Consider potential conflict between ductwork and existing piping and drains.

- **Clearance:** Usually 18 inches minimum for service access. The split unit should be at least 12 inches from the wall and accessible for service. Clearance from the wall may need to be more than 12 inches if the overhang from the roof blocks airflow.
- **Obstructions:** All air inlets and discharges on the outdoor unit must have unrestricted airflow for proper operation. Weeds, shrubs, trees, and any other obstructions must be cleared to maintain proper airflow.
- **Refrigerant line connections and drains:** The drains should not make watery marshes at the unit.
- **Mounting:** Consider the pad upon which the unit is to be mounted or the structure to be used for mounting. The unit should be installed above normal snow lines and where prevailing winds will not affect system operation. The pad must be level and sufficient for the permanent or long-term installation. Blocks and bricks alone should not be used to level a system.
- Follow all manufacturer specifications, instructions, and procedures.

Clearance

Review Questions
Section 9: System Installation

Name: ______________________________ **Date:** ________________

1. How much clearance should be in front of the indoor unit for servicing purposes?

2. What should be installed with an air handler in an attic?

3. What is a bi-product of an HFC refrigerant exposed to an open flame?

4. How much clearance should be left around the outdoor unit for servicing purposes?

5. When should both liquid and vapor lines be insulated?

6. When using a brazing torch on a refrigerant line, what should the technician also be using during this process to protect the system?

7. Why is it so important to level the air handler?

8. How should a horizontal air handler be mounted to the floor joists in some basements?

9. What guide is used to size wiring for the heat pump?

10. When installing the outdoor section of a split system in a location with snow, what precaution should be taken?

Objectives

Upon completion of this section, the participant will be able to:

1. describe the minimum requirements for system evacuation
2. list the steps in the triple evacuation method
3. understand the three charging methods; weight, superheat and subcooling

Evacuation

In the past, problems caused by poor evacuation techniques did not show up until sometime after the equipment was installed—usually after the compressor's warranty had expired. With industry's transition to HFC refrigerants and ester-based oils, it is detrimental to the equipment if proper installation techniques are not used. Air and moisture in these systems will cause major problems to develop much faster than equipment using other types of oils and refrigerants.

Ester-based oils absorb large amounts of moisture compared to other commonly used oils. This extra absorption can create more problems with the operation of heat pumps, especially in very cold climates. The only means of removing moisture from ester-based oils is by chemical reaction—using a liquid line filter drier.

When performing an evacuation, the best course of action is to install any refrigeration system to industry standards, using the proper tools and techniques for installation.

The following guide can be considered a **minimum** requirement:

1. Remove access valve cores with a core replacement tool.
2. Use a four-valve manifold gauge set with a 3/8- or 1/2-inch vacuum line.
3. Triple evacuate with a two-stage vacuum pump and use nitrogen to break the vacuum.
4. Use a micron gauge to evacuate the system to 500 microns.

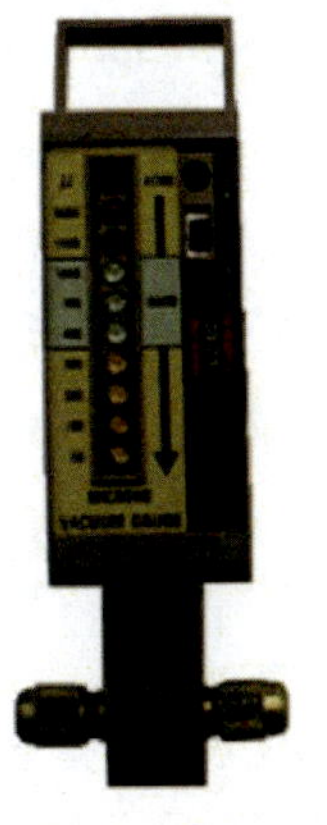

Micron Gauge

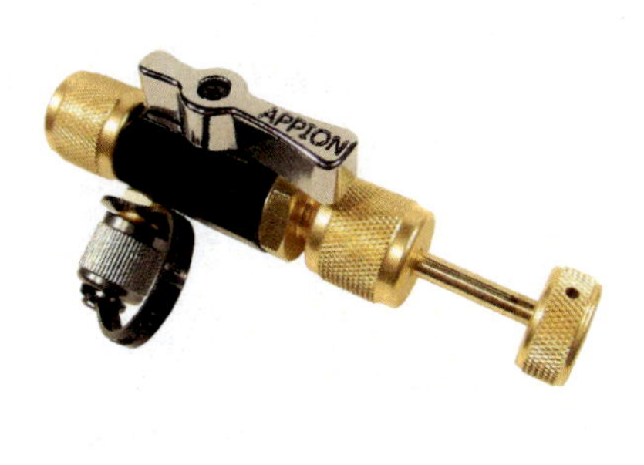

Valve Core Removal Tool

Performed properly, the triple evacuation method provides the same result as pulling a vacuum for twenty-four hours. The following steps should be used for triple evacuation:

1. Connect the manifold gauge set to the low and high side service valves.
2. Introduce nitrogen into the system until the pressure increases to 125 psig and check for leaks. Repair any leaks and recheck before proceeding.
3. Remove the nitrogen from the system: connect the micron gauge and vacuum pump.
4. Operate the vacuum pump, pulling from both the low and high side service valves until the micron gauge measures 1,500 microns.
5. Close the valve to the vacuum pump and reintroduce nitrogen into the system to a pressure of 1 or 2 psig.
6. After five minutes, bleed off the nitrogen pressure and pull the second vacuum until the micron gauge measures 1,500 microns.
7. Repeat step number 5.
8. After five minutes, bleed the nitrogen pressure and pull the third vacuum until the micron gauge measures 500 microns.

R-22 Manifold Gauge Set

Two-Stage Vacuum Pump

Charging

Weight method is the most accurate for any system. The installation instructions indicate the correct charge for the system and how much to add per foot of field installed line set.

System superheat is used for charging capillary tube or fixed orifice systems above 65°F outdoor temperature. This method cannot be used for charging a system that has a TEV metering device.

Subcooling is used for charging a TEV system.

NOTE: **Some systems will use both types of metering devices—a TEV for cooling and a fixed orifice for heating. Do not assume which is being used; always check the system and find out.**

Weight Method

The amount of refrigerant for each component must be added together when installing a split system to determine the total and correct charge.

Outdoor Unit + Indoor Coil + Line Set + Accessories = Total Charge

The amount of refrigerant per foot of line set listed below can be used when the manufacturer's chart is not available.

R-22:

5/16" OD LIQUID LINE - 3/4" suction = .46 oz. per foot
3/8" OD LIQUID LINE - 3/4" suction = .68 oz. per foot
3/8" OD LIQUID LINE - 7/8" suction = .70 oz. per foot

R-410a:

5/16" OD LIQUID LINE - 3/4" suction = .36 oz. per foot
3/8" OD LIQUID LINE - 3/4" suction = .55 to .62 oz. per foot
3/8" OD LIQUID LINE - 7/8" suction = .55 to .62 oz. per foot

New systems usually have enough refrigerant for the indoor unit, outdoor unit, and either a fifteen- or twenty-five-foot line set. With longer line sets, the amount of refrigerant must be calculated to insure that the charge is within one ounce (above or below). Check installation instructions; each manufacturer's requirements are different.

When using filter driers, the following chart can be used to approximate the amount of refrigerant to be added for a proper charge.

Dessicant	Ounces of Refrigerant by Weight					
Cubic Inch	R-134a	R-22	R-404	R-407c	R-410a	R-507
3	1.9	1.9	1.6	1.7	1.7	1.6
5	4.9	4.8	3.9	4.2	4.2	4.1
8	6.9	6.8	5.6	6.0	5.9	5.8
16	11.0	10.8	8.9	9.5	9.4	9.3
30	17.6	17.3	14.2	15.2	15.1	14.9
41	24.7	24.3	19.9	21.4	21.1	20.9

System Superheat Method

System superheat is refrigerant temperature measured from the inside of the evaporator to the suction inlet of the compressor. Superheat equals the suction line temperature at the compressor minus the saturated evaporator temperature (from PT chart). The required superheat is determined by indoor wet-bulb and outdoor dry-bulb temperatures. Superheat can be as low as 5°F or as high as 40°F, depending upon ambient conditions. System superheat is used for charging capillary tube or fixed orifice systems operating at outdoor temperatures above 65°F. This method cannot be used for charging a system utilizing a TEV metering device.

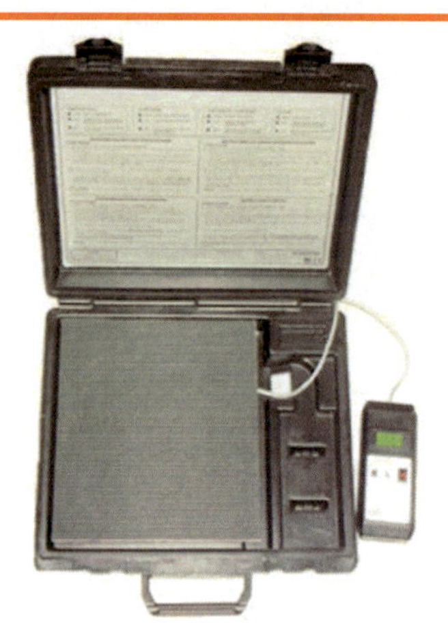

Refrigerant Charging Scale

NOTE: **Some systems will use both types of metering devices—the TXV for cooling and the fixed orifice for heating. Do not assume which is being used; always check the system.**

System superheat is used for charging fixed orifice or capillary tube metering devices on the cooling cycle only. The indoor temperature must be near normal (70-80°F) and the outdoor temperature above 70°F. As the outdoor temperature increases, the superheat goes down.

Example: 80°F ambient temperature and 64°F indoor wet-bulb temperature equals 15°F system superheat. (See system superheat chart on the opposite page.)

Measuring System Superheat

1. Find the saturated suction temperature (evaporator or suction line pressure changed to temperature, using a pressure/temperature chart).
2. Find the suction line temperature.
3. Suction line temperature minus saturated temperature equals system superheat.

NOTE: **Measured superheat should be within 5°F of the required amount.**

Subcooling Method

Subcooling is the additional cooling of refrigerant below its condensing temperature; it is the difference between the condensing temperature and the liquid line temperature.

SYSTEM SUPERHEAT CHARGING

Condenser Entering Air Temperature	Indoor Wet-Bulb Air Temperature: 52	54	56	58	60	62	64	66	68	70	72	74	76
65°F	6	10	13	16	19	21	24	27	30	33	36	38	41
70°F		7	10	13	16	19	21	24	27	30	33	36	39
75°F			6	9	12	15	[illegible]	21	24	28	31	34	37
80°F				5	8	12	15	18	21	25	28	31	35
85°F						8	[illegible]	15	19	22	26	30	33
90°F						5	9	13	16	20	24	27	31
95°F							6	10	14	18	22	25	29
100°F								8	12	15	20	23	27

To increase superheat remove refrigerant.	To decrease superheat add refrigerant.

System Superheat Chart - Most system superheat charts from various manufacturers are the same if the SEER of the system is about the same.

The manufacturer designs a specific amount of subcooling for the condensing unit. The subcooling temperature changes when there are significant changes in the outdoor ambient temperature.

The subcooling method is used for charging TEV systems. All systems must provide a certain amount of subcooling to prevent flash gas in the liquid line. System design, liquid line pressure drop, surrounding temperature, and evaporator temperature all have an effect on the amount of subcooling required.

Measuring Subcooling

Find the saturated condensing temperature (high side pressure changed to temperature, using a pressure/temperature chart).

Find the liquid line temperature leaving the unit. Saturated condensing temperature minus liquid line temperature equals subcooling.

NOTE: **Measured subcooling should be within 3°F of the required amount.**

Review Questions
Section 10: Refrigerant Evacuation and Charging

Name: ______________________________ **Date:** ________________

1. The __________ method is the most accurate charging method for any system.

2. The superheat method is used with what metering device(s)?

3. The __________ method is used for charging a Thermostatic Expansion Valve (TEV) system.

4. When performing the triple evacuation method, pull the third vacuum until the micron gauge reads __________ microns.

5. Why evacuate a refrigerant system?

6. Which oil attracts and holds moisture better; mineral or ester-based?

7. What should the technician do with the valve cores during the vacuum process?

8. Approximately how many ounces of R-410a would be needed for a drier 16 cubic inches?

9. What must be used to remove moisture from POE oil?

10. When measuring system superheat the calculated amount is 15°F and the chart indicates the required amount be 20°F. Should refrigerant be added or removed?

Lab Assignment 1

Section 10: Refrigerant Evacuation and Charging

Name: ______________________________ **Date:** ____________________

Objective: Calculate system refrigerant charge for a split system heat pump in ounces and lbs.

Directions: Use manufacturers literature to calculate recommended refrigerant charge using line set length, indoor and outdoor system. List refrigerant charge for each component.

NOTE: Perform the following task using all safety procedures!

System

Brand: ______________________________ Refrigerant type: ____________________

Outdoor
Model Number : ____________________

Indoor
Model Number : ____________________

Liquid line diameter: ____________________ Vapor line diameter: ____________________

Line set length: ____________________ Filter drier size: ____________________ cu. In.

Indoor unit charge: ____________________ Outdoor unit charge: ____________________

Line set charge: ____________________ Filter drier charge: ____________________

Total charge in ounces: ____________________

Total charge in pounds: ____________________

Objectives

Upon completion of this section, the participant will be able to:

1. describe checks included in pre-summer preventative maintenance
2. describe checks included in pre-winter preventative maintenance

Maintenance Checks

Heat pumps are usually dependable, working correctly for years; however, scheduled preventative maintenance before the heating and cooling seasons can still prove to be well worth the time and effort involved. Preventative maintenance assures an efficiently running system that reduces energy consumption. Many efficiency robbing problems can occur within a system and may not be noticed by the homeowner. A checklist should be used to insure that every item or function has been tested or checked for potential problems. A record should be made of the current thermostat setting and customer comments before starting any service or maintenance on the equipment. The customer and the technician should have a copy of the checklist, and the service company should keep the list on file for future reference.

Pre-summer preventative maintenance should include:

- Check indoor thermostat function, appearance, and level for operation, and make sure the cover is properly in place. If the thermostat is damaged, replace it with a new one. If it is digital with time set back, check the functions and operation.
- Check the return air grill and air filter. The air filter should be changed or cleaned (if it is a permanent filter) monthly for a residence; commercial systems may require a scheduled maintenance filter change more often.
- Check air vents, diffusers, and registers to insure proper unrestricted airflow. If the airflow is less than required, check the ducts and indoor coil for blockage or restrictions. Airflow can be hampered by loose fan blower belts, dirty coils, blocked vents, blocked ducting, blocked return air grill, or improper installation.
- Check the temperature split (temperature difference between the return and the supply register) after the system has stabilized.
- If the unit is a split system, check the entire blower assembly and fan motor to be sure the blower wheel is tight on the shaft.
- Check and clean the drain, drain pan, and condensate pump. Clean any mold and mildew deposits on or inside the air handler. If the problem is excessive, report it to the homeowner for further cleaning or replacements.
- Check for oil residue on the refrigerant coils and lines; this may indicate a refrigerant leak.
- Visually check the unit's ductwork system for damage and use duct tape or duct sealant on all air leaks and loose insulation.
- Check the outdoor unit for excessive fan noises or vibration, and proper operation.
- Check for loose wiring or damaged low voltage wiring; insulation damage may be caused by things like weed trimmers, or by animals.
- Check fuses or breakers for discoloration; this is caused by loose wiring or high amperes.
- Check the high/low voltage and amperage of system components and compare the readings to the manufacturer's nameplate.
- Make sure the outdoor unit is level and above the normal snow fall level.
- Plants, shrubs, privacy fences and porches or other coverings must not obstruct the outdoor unit's air intake or outlet vent paths.
- All panel screws and fasteners must be in place, or replaced by the technician if missing.
- All service valves must have caps to prevent refrigerant leaks (access valve cores are a secondary seal; the cap is the primary seal).
- If the system is not cooling properly, check the refrigerant charge. This should be done only if the system is not functioning properly. Every time gauges are used, the system loses refrigerant critical to the operation and system performance. If the charge needs to be adjusted, use the manufacturer's charts for exact performance pressures and temperatures for references. Lubricate all ports designed for that purpose.

Pre-winter preventative maintenance should include all the same basic checks, plus the following additional ones:

- The system should be energized in *heating* mode, and the heating cycle should be completely checked.
- Supplemental heaters should be tested or checked by the technician. Some heaters may have problems with open limits or open heat links due to airflow problems and hot spots.
- Check resistance heater wiring for burnt connections, loose terminals, corrosion, and discolored connections.
- Check the current draw of the heating elements (at 230 volts, the current should approximately 4.4 amps per kW).

Maintenance Form

____

Company Name

Date: _____/______/______ **Technician** ______________________________

Customer Name: ____ **Phone number:** ______ - _______-________

Address: ____ **Email:** ____________________________

City: __________________________ **State:** ________ **Zip Code:** __________

Make and model of heat pump ____

Customer comments: ____

____

Pre-summer Preventative Maintenance Check List

- ☐ Indoor thermostat condition and operation. Current temperature settings: ______°F Cooling, ______°F Heating
- ☐ Return air grill and air filter condition
- ☐ Air vents, diffusers, and registers airflow
- ☐ Airflow blocked or airflow problems
- ☐ Indoor unit voltage measured while operating: _________ volts
- ☐ Indoor unit amperage measured while operating: ________ amperes
- ☐ Low voltage measured at the indoor unit: __________ volts
- ☐ Indoor return air dry bulb temperature: __________°F
- ☐ Indoor supply air dry bulb temperature: __________°F
- ☐ Indoor wet bulb temperature: __________°F
- ☐ Supply/return air temperature split: __________°F
- ☐ Required temperature spilt from temperature split chart: _____°F
- ☐ Blower wheel and fans blades are tight on the shafts; both indoor and outdoor
- ☐ Motors have been lubricated (unless factory sealed)
- ☐ Drain pans and drain piping are clear of obstacles and debris
- ☐ Condensate pump has been cleaned and is working properly
- ☐ Air handler is free of mildew and mold, inside and outside
- ☐ All loose insulation and duct air leaks have been taped or sealed
- ☐ Outdoor fan motor has been checked for excessive noises, vibrations, and operations
- ☐ The unit has been checked for loose wiring and for damaged low voltage wiring
- ☐ Fuses and breakers have been checked for discoloration and potential problems
- ☐ Line voltage measured at the outdoor unit during operation: _______ volts
- ☐ Amperage measured at the outdoor unit during operation: ______ amps
- ☐ Low voltage measured at the outdoor unit: __________ volts
- ☐ Measured outdoor dry bulb temperature: ___________°F
- ☐ Measured outdoor unit leaving air temperature: _________°F
- ☐ Crankcase heater is functioning: _________ amperes
- ☐ All capacitors are visually okay (not enlarged or shorted)
- ☐ Outdoor unit is level and unobstructed by scrub, trees, bushes, etc
- ☐ All panels are in place and have screws in every screw hole
- ☐ All valves have valve caps in place

Pre-Winter Preventative Maintenance Checklist

- ☐ System has been checked in Heating mode and the heating cycle has been completely checked out
- ☐ Supplemental heaters have been tested or checked by the technician
- ☐ Heater's wiring has no burnt connections, loose terminals, corrosion or discolored connections
- ☐ Amperage measured at indoor unit while operating in Emergency Heat Mode: _________ amperes
- ☐ System defrost control operation tested

Comments/Findings ____

____

____

____

Review Questions
Section 11: Preventive Maintenance

Name: ______________________ **Date:** ______________

1. __________ assures an efficiently running system that reduces energy consumption.

2. The air filter in a residence should be changed or cleaned __________.

3. A __________ should be used to ensure that every item or function has been tested or checked for potential problems.

4. If the system is not cooling properly, check the refrigerant charge, however, this should be done only if the system is not functioning properly as every time gauges are used, the system loses __________ critical to the operation and system performance.

5. When should the customer have preventive maintenance preformed (PM)?

6. Does a PM include checking the thermostats? Why or why not?

Objectives

Upon completion of this section, the participant will be able to:

1. understand the importance of checking the refrigerant charge
2. list 5 main questions to ask before servicing a unit
3. describe the methods used to calculate the net cooling capacity
4. understand how to troubleshoot a compressor
5. identify troubleshooting procedures for refrigerant components
6. troubleshoot electrical and control components utilizing the wiring diagram
7. effectively use a troubleshooting flowchart

Refrigerant Charge

Manufacturer's data and specifications should be used for servicing and troubleshooting a specific unit. Higher efficiency heat pumps have different requirements and specifications. There is no longer a set charging formula, such as ambient plus 30°F converted to pressure. All suggested readings are based on the unit operating in a *normal* environment. If the unit is operating out of its designed range, the pressure and temperature measurements will also be out of range. The suction pressure may be greater because the load is greater, not because the unit is malfunctioning. It is almost impossible to tell how much refrigerant is in a system that has a suction accumulator while operating in heating mode. Plus or minus half an ounce of refrigerant charge can make a difference in capacity. The only way to charge a system without guessing is to weigh the refrigerant charge into the system. System superheat and subcooling do not provide the perfect charge; these methods depend upon airflow and temperature being within the proper range at the time that the system is charged.

In the event of a mechanical problem with a heat pump system, troubleshooting can be minimized if the refrigerant charge is weighed in. A faulty check valve, metering device, reversing valve or compressor can change operating pressures, making it seem like the system is properly charged. When the charge is weighed in and the system is not performing to specifications, a lot of time can be saved if the work of verifying the charge can be skipped. This allows the technician to focus on checking the mechanical components for a problem.

General Troubleshooting Tips

Before attempting to service any heat pump, time should always be taken to understand the sequence of operations for the system being addressed. Without a proper understanding of how an individual unit operates, it is nearly impossible to service the unit using a systematic approach.

Having the answers to the following questions before attempting to service any unit will help pinpoint problems quickly and easily:

- Does the problem occur during heating or cooling?
- Is the reversing valve energized in heating or cooling mode?
- Does the problem occur due to the unit not defrosting properly?
- What type of defrost system does this heat pump use?
- Is the problem really caused by a malfunction, or is the unit functioning normally?

While the last question may seem strange at first, it is a fact that some customers will call for service even if the unit is operating normally. For example, sometimes a customer may hear a strange sound (shifting of the reversing valve), see what looks like smoke coming from the outdoor vent (usually fog from the coil during defrost), or feel cold air coming from the vents (could be a normal condition during the defrost cycle).

Do *not* put gauges on the system unless absolutely necessary; Refrigerant charge is critical for heat pumps and liquid refrigerant loss from improperly removing the manifold gauge hose from the high-side liquid line service port is enough to change the operating efficiency.

Net Cooling Capacity

This is a simple check to measure the total sensible heat rejected by the outdoor air. It can be used on a heat pump or a cooling system.

When performing the calculations, the outdoor fan CFM must come from the manufacturer's data. The design of the fan discharge areas from one system to another varies too much to use an average area for CFM calculations.

The Btu capacity of a system can be calculated by using the indoor airflow; this is the most difficult method. Calculating total Btu capacity from the evaporator involves measuring airflow and dry- and wet-bulb temperatures entering and leaving the evaporator, then plotting the measurements on a psychometric chart to find latent and sensible heat in Btu. Heat from the condensing coil or outdoor unit is sensible heat rejected to outdoor air plus the heat from the electrical loads (compressor and fan). The net Btu capacity can be calculated without using a psychometric chart.

The following formulas are used with the measurements from the outdoor unit:

Sensible Heat =
1.08 (1.1) x Outdoor Fan CFM (from Manufacturer Data)
x Average Condenser Air Temperature Difference
(an average of temperature measurements must be used)

Electrical Motor Heat =
Volts (line input voltage) x Amps (compressor and fan motor) x 3.41 Btu per Watt x Power Factor (average is 0.9; higher efficiency equipment will be a slightly higher number)

System Btu capacity =
Sensible heat Btu - Btu of motor heat

Because test instruments used in the field can sometimes be inaccurate there can be an error of plus or minus 2,000 Btu. This is not a large number compared to the total Btu of the average heat pump system.

Refrigerant Charge

Superheat

Superheat can be used to check the refrigerant charge, as previously shown, only in the cooling cycle on a system with a capillary tube or fixed orifice. Indoor air CFM must be in the proper range and indoor wet-bulb and outdoor dry-bulb temperatures must be accurate.

Subcooling

Subcooling is the difference between the liquid line temperature at the condenser outlet and the condensing temperature. Subcooling measures how much additional cooling the liquid refrigerant receives after it has condensed into liquid. Subcooling should be checked on systems with a TEV and the amount of sub-cooling based upon the manufacturer's specifications.

Evaporator Temperature Split Check

The temperature split across the evaporator is a design factor based on the sensible heat ratio, which is calculated using the heat load of the structure and the required CFM of air. The system runs with the required temperature split only at the design conditions. As the indoor relative humidity increases, the temperature split decreases.

Troubleshooting

Compressors

Troubleshooting a compressor is much easier if the technician understands all probable causes of compressor failure. In many instances, a compressor is condemned by the technician, and it is later found that it was not the compressor that failed, but something else in the electrical or refrigerant circuit. The following guide will help to reduce this problem. Many compressor malfunctions are due to *system overcharging*. This is perhaps the biggest problem with compressors and a major cause of compressor failure. To make sure things are done correctly, a technician must follow the manufacturer's charge procedures and weigh-in the refrigerant. On a split system, refrigerant charge must be calculated for the size of the filter drier and length of the refrigerant lines and any accessories. This calculated amount of refrigerant must be added to the system's listed total data plate charge.

Problems that will be covered in the following pages:

- Compressor is hot, does not run, and outdoor fan is running (internal overload).
- Compressor runs, but does not pump properly (broken valves in compressor).
- Compressor blows fuses or kicks the breaker very quickly (compressor windings shorted or grounded).
- Compressor hums, draws high amperes, or will not start (faulty run capacitor or start capacitor).
- Compressor does not start or run, blows fuses, or kicks the breaker (locked rotor).
- Compressor does not start or run (open compressor windings).

Compressor is hot, does not run, and the outdoor fan is running:

If the system is running in cooling mode and the service call is "no cooling," first check the outdoor unit for problems. Check air temperature out of the outdoor unit to confirm if the system is moving heat. This indicates whether or not the compressor is operating. If the outdoor fan is running, this confirms that there is low and high voltage to the unit.

Check the temperature of the compressor, being careful not to get burned. If the temperature is very high, the internal overload may be open. An open internal overload can be verified by using an ohmmeter. With the power supply turned off, check the resistance of the compressor from the run and start terminals and then each to common. Continuity between only run and start means the internal overload is open.

Check all wiring to the compressor. Check the run capacitor and start capacitor, if used. If all electrical components are correct, check the refrigerant at the suction valve to confirm that there is refrigerant in the system. A system with a low charge will ice up and one that is almost empty will cause the compressor to overheat.

If refrigerant has leaked out of the system, find and repair the leak. *Never* just add refrigerant.

Compressor runs, but does not pump properly:

If the system runs in cooling mode but the refrigerant has a very high *suction* pressure and a very low *liquid line* pressure check the suction line temperatures going in and out of the reversing valve. If the valve inlet suction line to the compressor and the line from the coil serving as the evaporator have a significant temperature difference (5 or more degrees) across the reversing valve, then the valve is faulty, not the compressor, and should be replaced due to internal leakage.

If the reversing valve is functioning correctly, the check valves or metering device may be allowing refrigerant to bypass. Check for lower than normal superheat. Reverse the system from heating to cooling, or cooling to heating. If pressures return to normal, there is a leaking check valve.

With all other components ruled out, check the running amperage of the compressor. Lower than normal ampere draw of the compressor usually indicates a compressor with faulty valves or other mechanical problems.

Sometimes if the valve is broken, the compressor makes a faint ringing sound while running. Compressor valves usually break because of flood-back or severe overcharging of refrigerant. Low indoor airflow can also cause liquid flood-back; this is the most common cause of compressor valve failure.

Compressor blows fuses or kicks the breaker very quickly:

If the breaker is in the tripped position or the fuses are blown, this is a serious situation that can be dangerous. Do not try to reset the breaker until a full search for all electrical malfunctions, broken wires, loose wiring connections, and discolored wiring lug connections has been conducted.

Warning:
Applying power to a compressor with grounded or burnt windings may cause the terminal block to blow out, releasing refrigerant and oil.

Check the compressor for a grounded condition by first removing the start, run, and common connections and measuring the resistance of the windings with a good ohmmeter. Measure each of the terminals to a good copper tubing connection on the compressor. If any of the compressor terminals indicate continuity to the copper, the compressor is grounded and must be replaced. If the start or run windings measure differently than the manufacturer's specifications, the compressor may be shorted. If the compressor's windings measure *infinity* or if the compressor has an open winding condition between start and run, the compressor must be replaced.

Anytime a compressor is in a grounded or shorted condition, the system may have burned oil and there may be acid inside the system and/or compressor. This must be cleaned before a new compressor is installed. New liquid-line and suction-line driers must be installed in order to clean up the system. If an acid test of the oil reads positive, a liquid line drier rated for acid clean-up must be used.

Grounded and shorted conditions can be caused by acid in the system, excessive non-condensables in the system or refrigerant restrictions which can deprive the compressor of vital cooling from suction gases.

Compressor hums, draws high amperes, and will not start:

If the compressor does not run, draws high amperes and will not start, always look first for obvious causes such as loose or burnt wiring. If the run capacitor is "bulged out," it may be quickly diagnosed. A bulged capacitor is usually a shorted capacitor. Check the capacitor with the proper meter and replace it with the manufacturer's recommended replacement.

Sometimes the compressor may have burnt terminals due to loose push-on terminals. Clean the terminals and replace any burnt wiring and push-on terminals.

Some heat pump compressors have a PTC (Positive Temperature Coefficient) thermistor starting device connected across the terminals at the run capacitor. This component may be used with or without a start capacitor. Check the PTC to make sure it is in working condition. The PTC is used to help soft-start the compressor. At temperatures between 70 and 95°F, the cold resistance of the PTC should be ten to twenty ohms cold. Once the PTC is de-energized from the circuit, it takes five minutes for it to cool down and for the resistance to drop back to normal.

PTC Thermistor

Check the contactor's contacts for any pitting, sticking or burning. Do not clean or repair the contacts: replace the contactor.

If the compressor has a start capacitor and starting relay, make sure they are working properly.

Compressor does not start or run, blows fuses, or kicks the breaker:

This condition is very similar to the previous one, but it has a different cause. A compressor with this problem has a locked rotor condition and will pull the Lock Rotor Amperes as shown on the nameplate. The best way to determine if there is a locked rotor is to go through the same procedures listed in the previous example: check all components, wiring, windings, and capacitors; if the compressor will not start, a locked rotor condition exists and the compressor must be replaced. A locked rotor can be the result of a seized shaft or small chips of copper, slag, or foreign materials inside the compressor bearing.

Compressor does not start or run:

When a compressor has an open winding, it may try to start on the remaining closed windings, but will not run. Find the open winding by checking the common terminals to start and run. If the open circuit exists only between the common and the other two terminals, check for an open internal overload. A very warm or hot compressor indicates an open internal overload; the measurements should confirm this diagnosis. However, if the compressor is cold and the open is found in the windings, the compressor must be replaced.

Refrigerant Components

Evaporator Temperature Split Check

The temperature split across the evaporator is a design factor based on the sensible heat ratio and is calculated using the heat load of the structure and the required CFM of air. The system runs with the required temperature split *only at design conditions.* As indoor relative humidity increases, the temperature split decreases. The following chart is used to calculate the temperature split for given conditions.

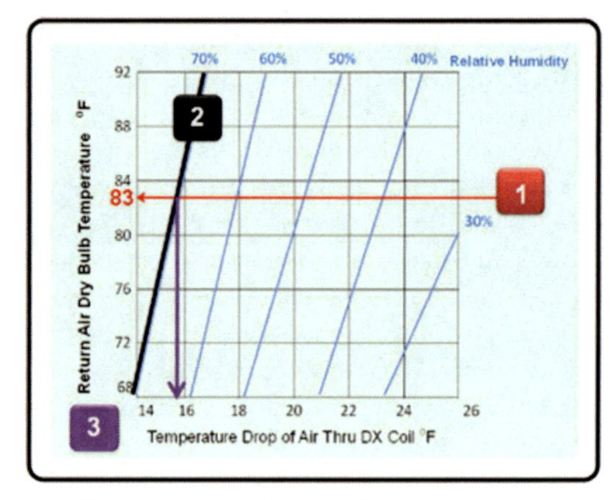

Line 1: Return Air Dry-bulb Temperature

Line 2: Indoor Relative Humidity

Line 3: Temperature Drop Through Evaporator

Electrical Components

The low-voltage controls in a heat pump system consist of many different components. Low voltage is supplied by a step-down transformer. Most heat pumps are either single-phase, 230-volt or three-phase 208/230, 440-volt, as with light commercial equipment.

Step-Down Transformer

Low voltage makes the system's control assembly safer, cheaper, and easier to manage. Almost all heat pump thermostats operate on low voltage, which can be checked safely with an AC voltmeter.

One easy method for checking many heat pump components is to go directly to the indoor thermostat and place the blower fan motor control in the On position. If the system is a split type, the transformer is usually found in the indoor section. If the blower comes on as it should (some are time-delayed) the technician knows there is high and low voltage available and the indoor fan motor is working.

Some manufacturers install a time-delay device in the indoor air handler which turns the indoor blower motor on and off. This device allows the system to come on in the cooling and heating modes, without the blower running, for about a minute. It also delays the blower's Off cycle.

Note: The resistance heating coils cannot come on without the blower running.

If the indoor section is operating properly, the problem is most likely with the outdoor unit.

On a split system, the outdoor section has the following low-voltage components:

Contactor

The contactor turns the compressor and outdoor fan motor on. The contactor has one low-voltage lead directly from the transformer (common) and the other lead is from the "Y" or yellow on the indoor thermostat, which is fed by "R." When the contactor has twenty-four VAC present at the contactor's coil, it should immediately snap the contacts closed and the compressor/outdoor fan should come on, *unless the outdoor unit is time-delayed.*

Single-Pole Contactor

Some units have a time-delay control that delays the contactor's action for a few minutes. If the contactor is not energized within five minutes, the time-delay control itself is most likely faulty, and must be replaced. This time-delay control is used by some manufacturers to delay the compressor, in order to prevent repeated immediate cycling on and off.

It should be noted that not all contactors have a low-voltage coil; some are high voltage on commercial equipment.

Reversing Valve

The reversing valve is low-voltage-operated for many units, but can also be high-voltage. For safety purposes, *do not* assume it is low-voltage.

A low-voltage reversing-valve relay or defrost relay can also be used for control purposes if the reversing-valve solenoid is operated by high voltage. In some units, the reversing-valve solenoid is energized during the heating cycle, but in most it is energized during the cooling cycle.

Defrost Control

The defrost control is usually electronic. It has an adjustable time setting which allows the control to enter defrost mode when needed. Most of the time, when this control fails it will not operate when the test terminals are manually connected. The defrost termination thermostat that is connected to the outdoor coil prevents the defrost control from operating if it does not close or open at the designed temperature range. If it takes the outdoor coil a few days to ice over, the defrost control must be set for a shorter time interval; for example, 60 to 30 minutes.

High-Voltage Controls

Troubleshooting high-voltage controls, such as a contactor or motor control, can easily be accomplished by using a good meter and some common sense safety procedures:

- Always use eye protection and follow all safety procedures.
- Remember, the technician is only as good as the testing equipment being used. Using reliable test equipment is the first step in being safe.
- Be sure that the voltage present is the same as the manufacturer's requirement.

When troubleshooting a heat pump's high voltage controls, first look at the ladder diagram for the system to determine the sequence of operations and which high-voltage components should be energized or have power. The presence of voltage can be determined by placing one lead from the meter to a known voltage source (L1), and moving the other lead from place to place. If voltage is not present, look at the diagram and follow the circuit back to the source (L2). When voltage is found, you will have found the break in the circuit.

Breakers and Fuses

Breakers and fuses are used to protect the high-voltage power supply to the system and should be mounted on or within sight of the equipment. Outdoor units must have a disconnect or means of disconnect. Check for discolored fuses and breakers, which can be caused by loose wiring and high amperage heating.

Fuse and breaker sizes are matched with the correct wiring sizes as required by the National Electric Code (NEC). Local codes can be more stringent, but *not* less than NEC requirements.

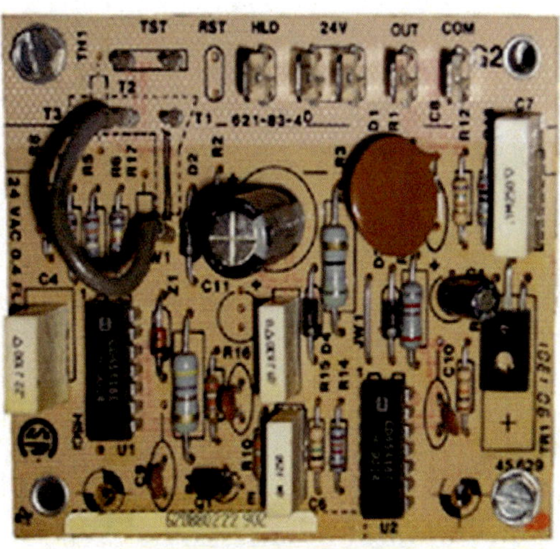

Defrost Control

Troubleshooting

Diagram 1 on the following page shows the meter reading 120 volts across the contacts. The red lead is connected to the L1 side of the contact and the black lead to the L2 side. The black lead is measuring voltage from L2, *through* the coil or load. With the contact open, the meter is in series with the load and completes the circuit. It measures the voltage because it has high resistance (approximately 20,000 ohms), which creates the largest voltage drop.

If the contacts were closed, the load would be energized and the meter would show a reading of zero volts. The electrons would take the path of least resistance through the contacts, bypassing the voltmeter.

This is only one of the checks that should be done, and it should not be used to determine whether main power to the circuit is off or on. If the load is defective or open, the voltmeter with not measure a voltage across the contact.

Diagram 2 shows the red lead on the line side of the contact and the black lead on the load side, as in Diagram 1. In this case, however, the meter is showing a reading of zero volts. Even though the meter indicates zero voltage, the circuit is energized with 120 volts going to the load. The electrical current is taking the path of least resistance through the contacts, bypassing the meter. If the voltmeter displays any voltage across closed contacts, it means that the contacts are defective or open.

Diagram 3 indicates zero voltage with the contact open. The red lead is connected to the line one side of the contact and the black lead to the load side. The voltmeter cannot measure the voltage due to the burned open spot indicated on the load. The voltmeter is in series with the load, but a complete path cannot be made to the L2 side of the circuit.

Diagram 4 indicates 120 volts with the test leads placed on the L1 and L2 sides of the load. This is the same as checking the power source to the circuit.

Troubleshooting a system that has an electrical problem can be both challenging and rewarding.

- Always remember that *the power may be on.*
- Always use eye protection and follow proper safety procedures.
- Some safety procedures may include keeping one hand in your pocket or away from the equipment at all times. This is to help prevent electrocution.

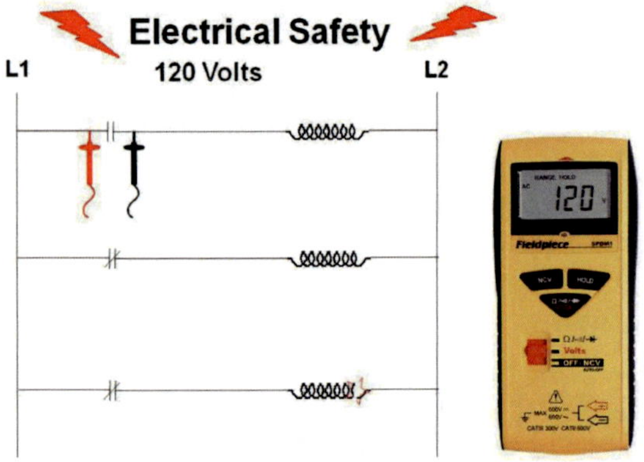

Electrical Safety
120 Volts
L1
L2
120

Diagram 1

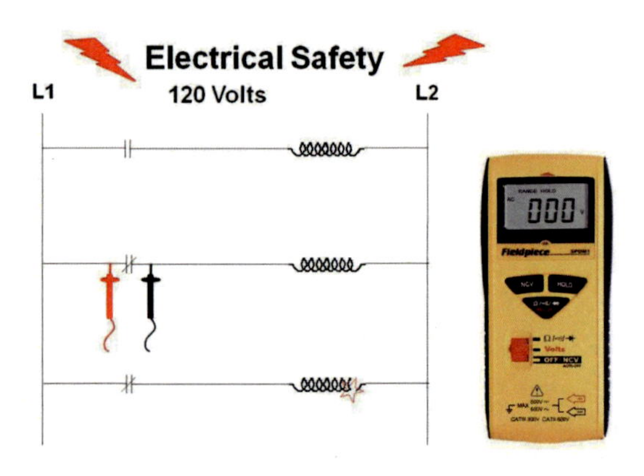

Diagram 2

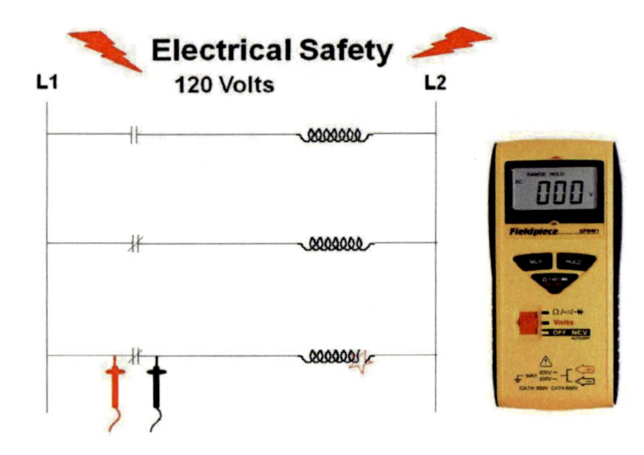

Diagram 3

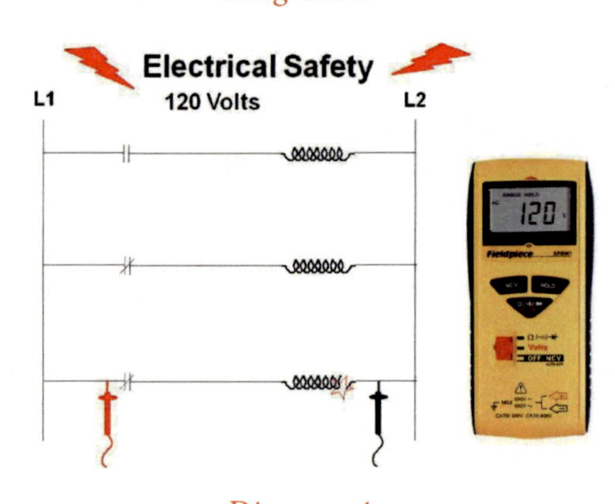

Diagram 4

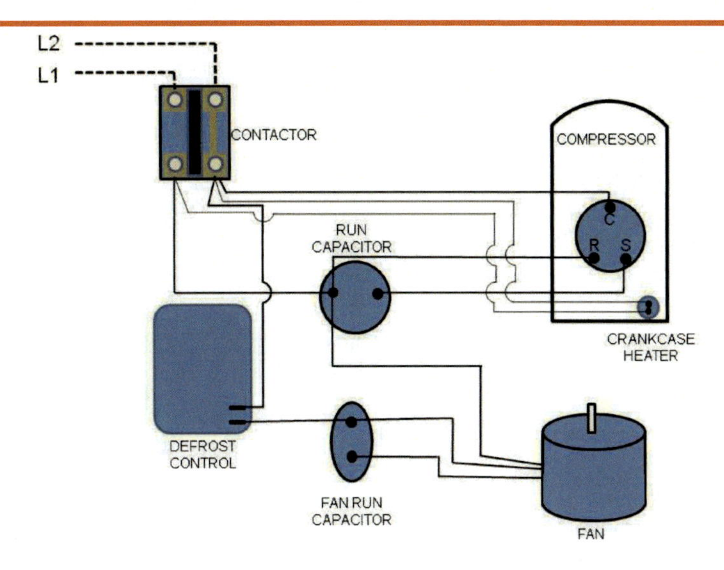

Pictoral Diagram

- Keep all surrounding equipment away from the electrical panel box or exposed electrical equipment while working in it.
- All jewelry worn by the technician *must* be removed before entering a hot electrical panel or control box. Rings and other jewelry can be extremely hazardous to the technician.

Wiring Diagrams

Shown below is a typical heat pump diagram. Notice that the high-voltage areas are located at the top of this ladder wiring diagram.

Near the center of the diagram is the step-down transformer, which supplies twenty-four volts to the low-voltage section. This section shows the indoor thermostat and the low-voltage controls used to turn the system on in cooling or heating mode.

Notice the defrost control, which is the solid-state type.

This diagram is easy to follow when troubleshooting the system electrically. It also makes it easy to see which control does what, and when. Ladder wiring diagrams are always shown in the non-energized position, unless labeled otherwise.

Troubleshooting Using the Ladder Diagram

One of the first steps in troubleshooting a system is learning the legend of symbols. A technician must learn the manufacturer's symbols; there are almost always some symbols that vary with each manufacturer. Take time to find the symbol for each component and look for the contacts belonging to that control.

For example, at the right of the wiring diagram, find the compressor and outdoor fan. A quick glance shows that the CH (crankcase heater) is energized all the time, from L1 and L2. The compressor and outdoor fan should come on if the CR (control relay) contacts are closed. Another quick look in the low-voltage section shows us the location of the control relay. To the right and in series with the control relay is a high-pressure control.

We can see that if the HP control were open, the CR relay coil would not have power, and the CR contacts located in the high-voltage circuit would not be closed, and therefore neither the compressor nor the outdoor fan motor would be operating.

Pictorials

Pictorial diagrams show all electrical components as they appear in the control panel. For instance, if the diagram shows the defrost relay between the transformer and the control relay, the panel would also look like this. Pictorials are useful for finding various colored or numbered wires when the technician is tracking them.

Using only a pictorial for troubleshooting purposes can be very confusing; however, the technician can use the ladder and pictorial *together* to expedite the troubleshooting process.

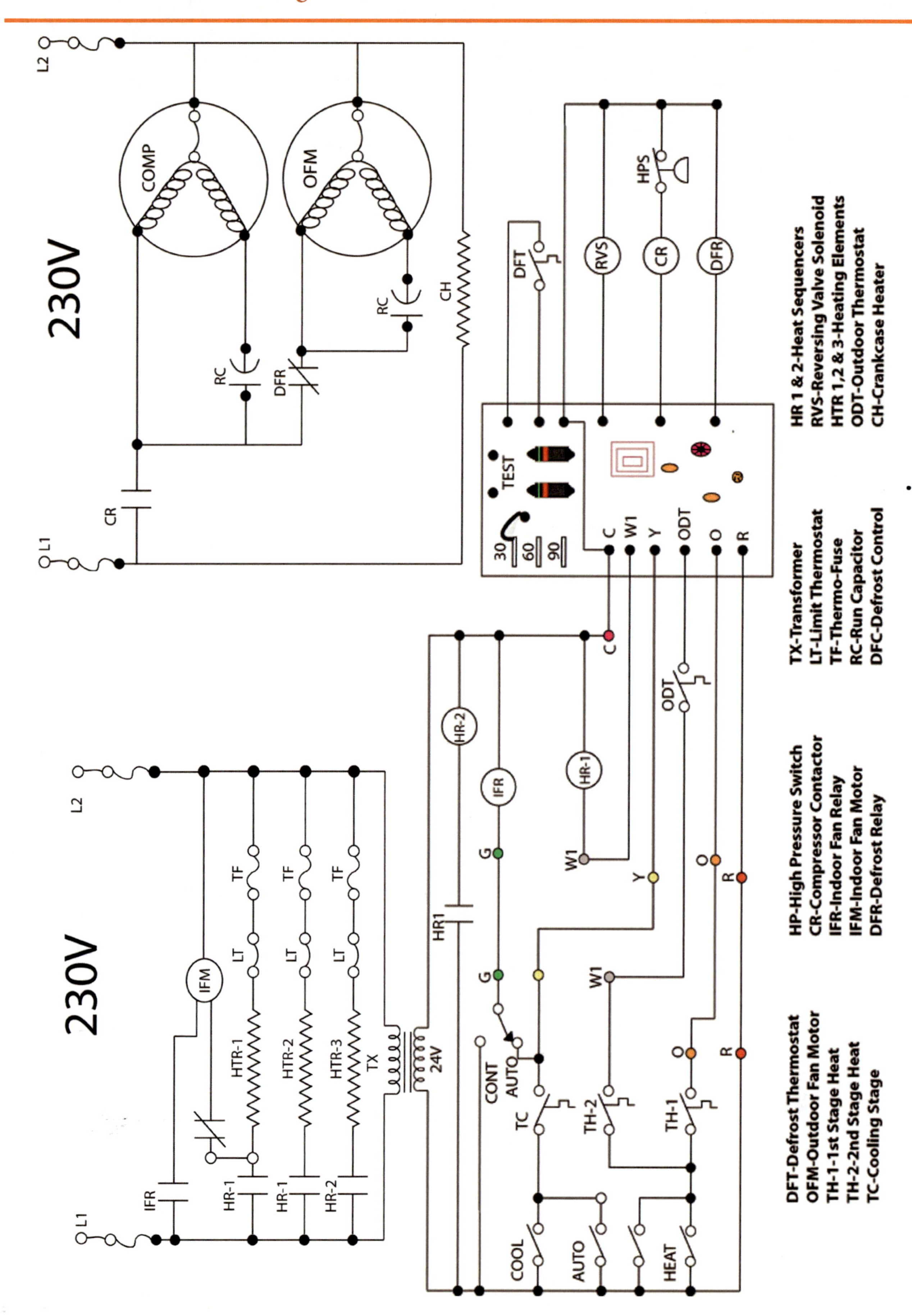
230V
230V
DFT-Defrost Thermostat
OFM-Outdoor Fan Motor
TH-1-1st Stage Heat
TH-2-2nd Stage Heat
TC-Cooling Stage
HP-High Pressure Switch
CR-Compressor Contactor
IFR-Indoor Fan Relay
IFM-Indoor Fan Motor
DFR-Defrost Relay
TX-Transformer
LT-Limit Thermostat
TF-Thermo-Fuse
RC-Run Capacitor
DFC-Defrost Control
HR 1 & 2-Heat Sequencers
RVS-Reversing Valve Solenoid
HTR 1,2 & 3-Heating Elements
ODT-Outdoor Thermostat
CH-Crankcase Heater

DETERMINE UNIT NET COOLING CAPACITY WORKSHEET

Student Name:____ **Date:**____________________

PROCEDURE:

1. **DETERMINE CONDENSER FAN CFM:**
   (From Manufacturer's Literature) __________**CFM**

2. **MEASURE CONDENSER ENTERING AIR TEMPERATURE:** __________ **F.**

3. **MEASURE CONDENSER LEAVING AIR TEMPERATURE:** __________**F.**

4. **SUBTRACT LINE 2 FROM LINE 3 TO OBTAIN TEMPERATURE SPLIT:** __________ **ΔT.**

5. **MEASURE VOLTS AND TOTAL AMPS AT**
   CONDENSING UNIT THEN MULTIPLY
   V x A x 3.41 x POWER FACTOR OF UNIT.
   THE ANSWER IS MOTOR HEAT IN BTUH: **RECORD** __________ **BTUH**

6. **UNIT NETCAPACITY= 1.08 x CFM x TD**
   MINUS MOTOR HEAT (STEP 5): **RECORD ANSWER**_______________ **BTUH**
   (Should be ± 2,000 Btuh of system capacity)

Heat Pump Troubleshooting Flow Chart #1

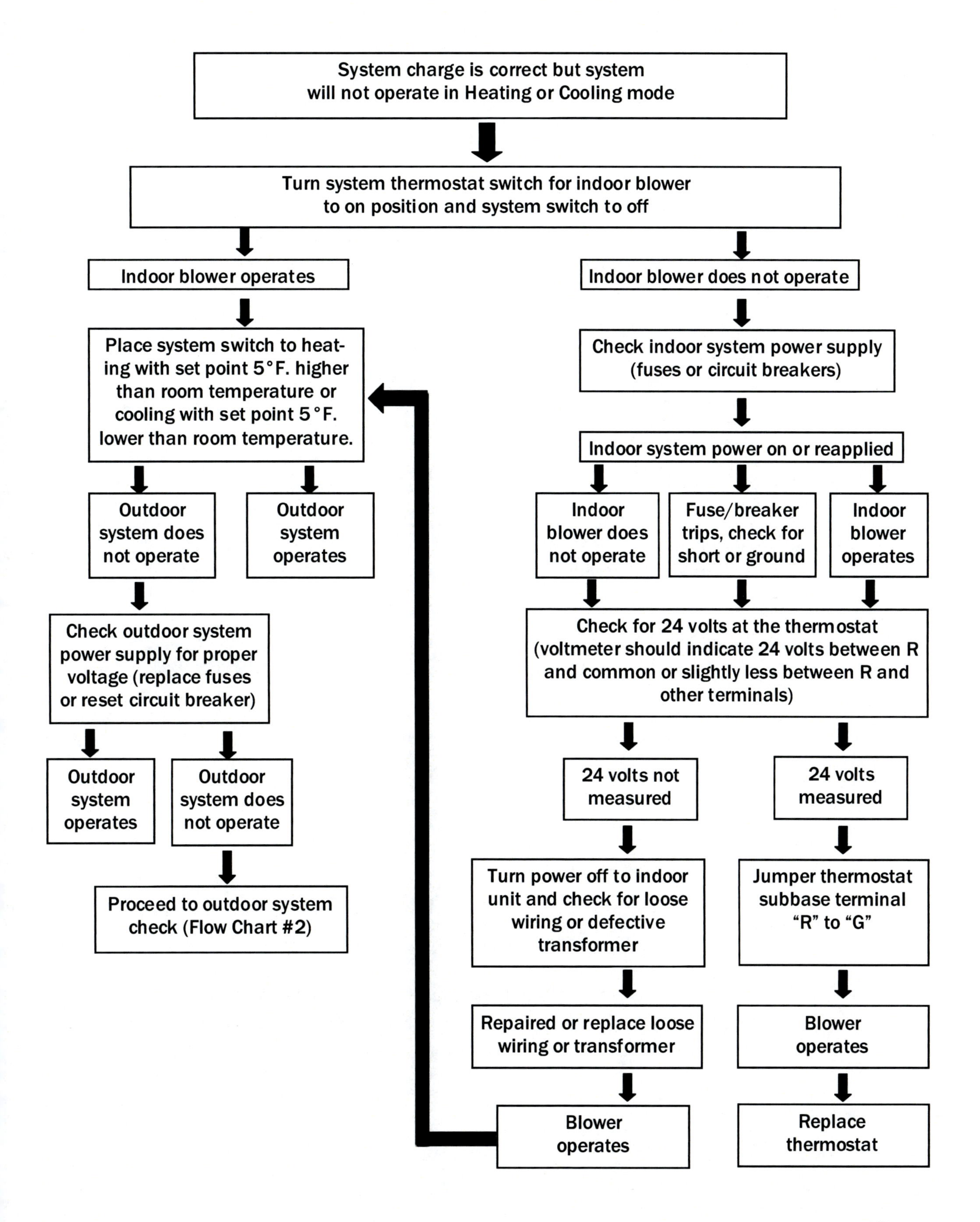

Heat Pump Troubleshooting Flow Chart #2

Outdoor section will not operate on Heat or cooling

⬇

Place system switch to heating with set point 5°F. higher than room temperature or cooling with set point 5°F. lower than room temperature

⬇

Check outdoor system power supply for proper voltage. (before replacing fuses or reset ting the circuit breaker, check for system circuit grounding with an ohmmeter)
Caution reapplying power to a grounded or burned compressor winding may cause the compressor terminal block to blowout releasing refrigerant and oil. If there is not a grounded or shorted condition, proceed with troubleshooting.

⬇ Outdoor system operates

⬇ Outdoor system does not operate

⬇

Check for 24 volts at the compressor contactor coil, the defrost control board and reversing valve terminal if system is set to cooling

⬇ 24 volts is not measured in either of the locations
⬇ Turn power off to indoor and outdoor systems and check for loose or defective control wiring

⬇ 24 volts measured at the defrost control but not the contactor coil
⬇ Check for open safety switch (low / high pressure, discharge / compressor temperature)
⬇ Reset or replace safety switch

⬇ 24 volts measured at the compressor contactor coil
⬇ Replace contactor

NOTE: Some equipment may have a solid-state control board that controls all functions of the outdoor system. If there is 24 volts to the appropriate terminals from the thermostat and all safety switches are closed, the board is probably defective.

Heat Pump Troubleshooting Flow Chart #3

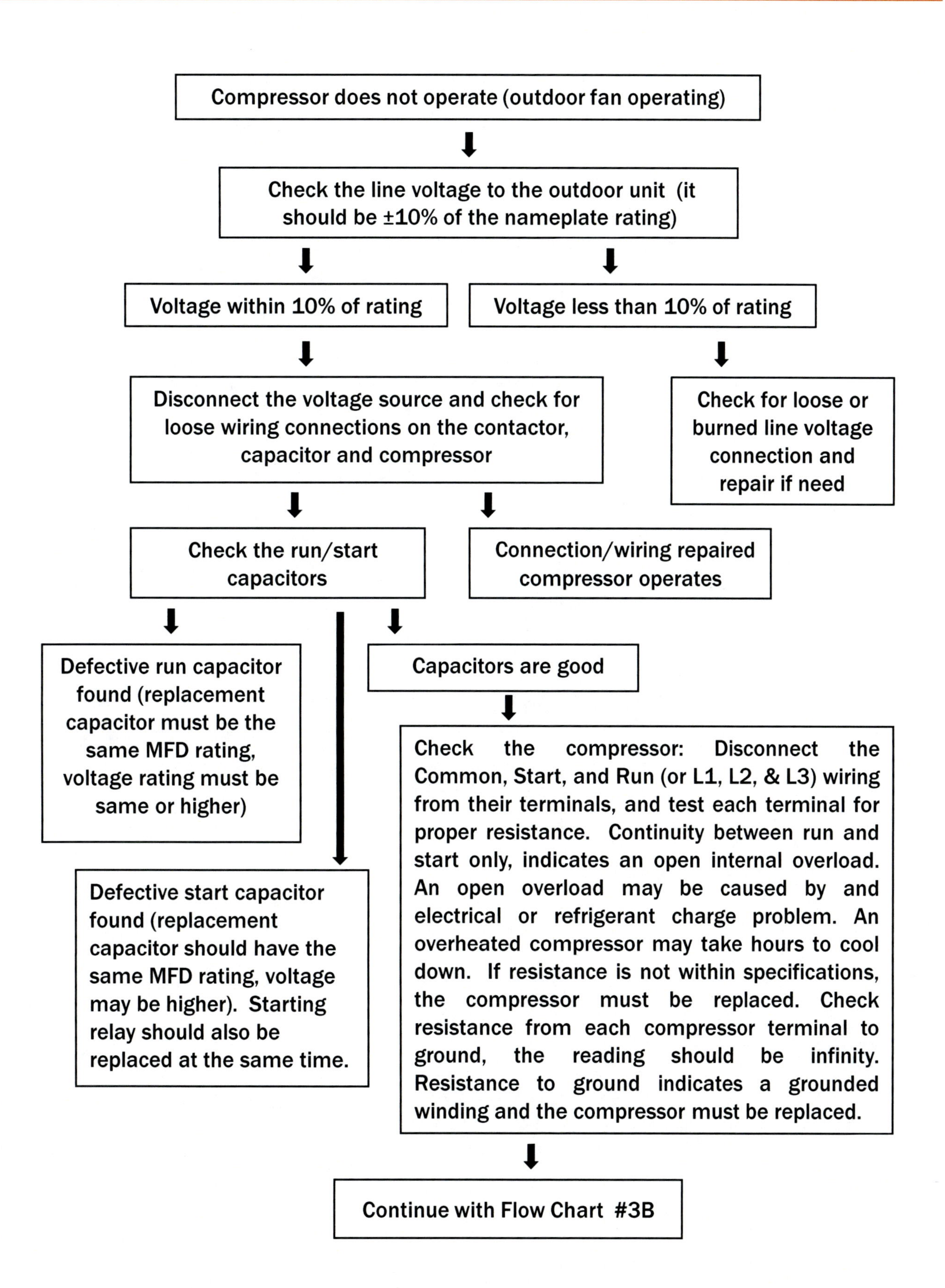

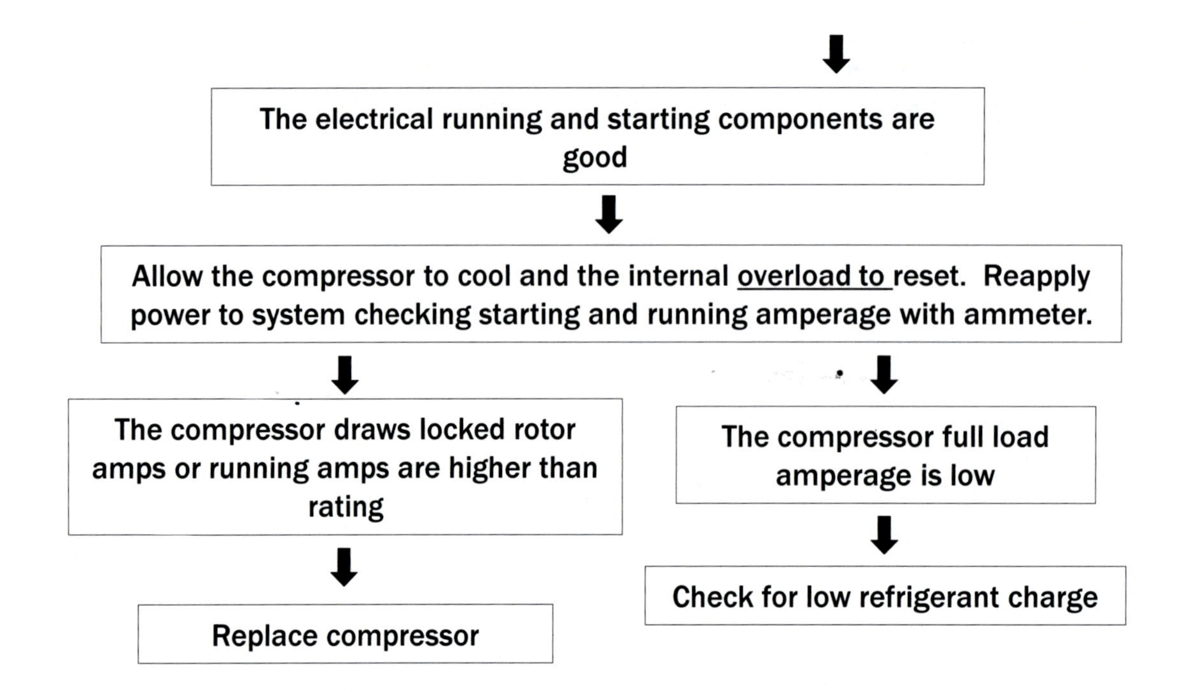

NOTE: The cause of the failed component should be found. There is always a reason for a component failure.

Heat Pump Troubleshooting Flow Chart #4

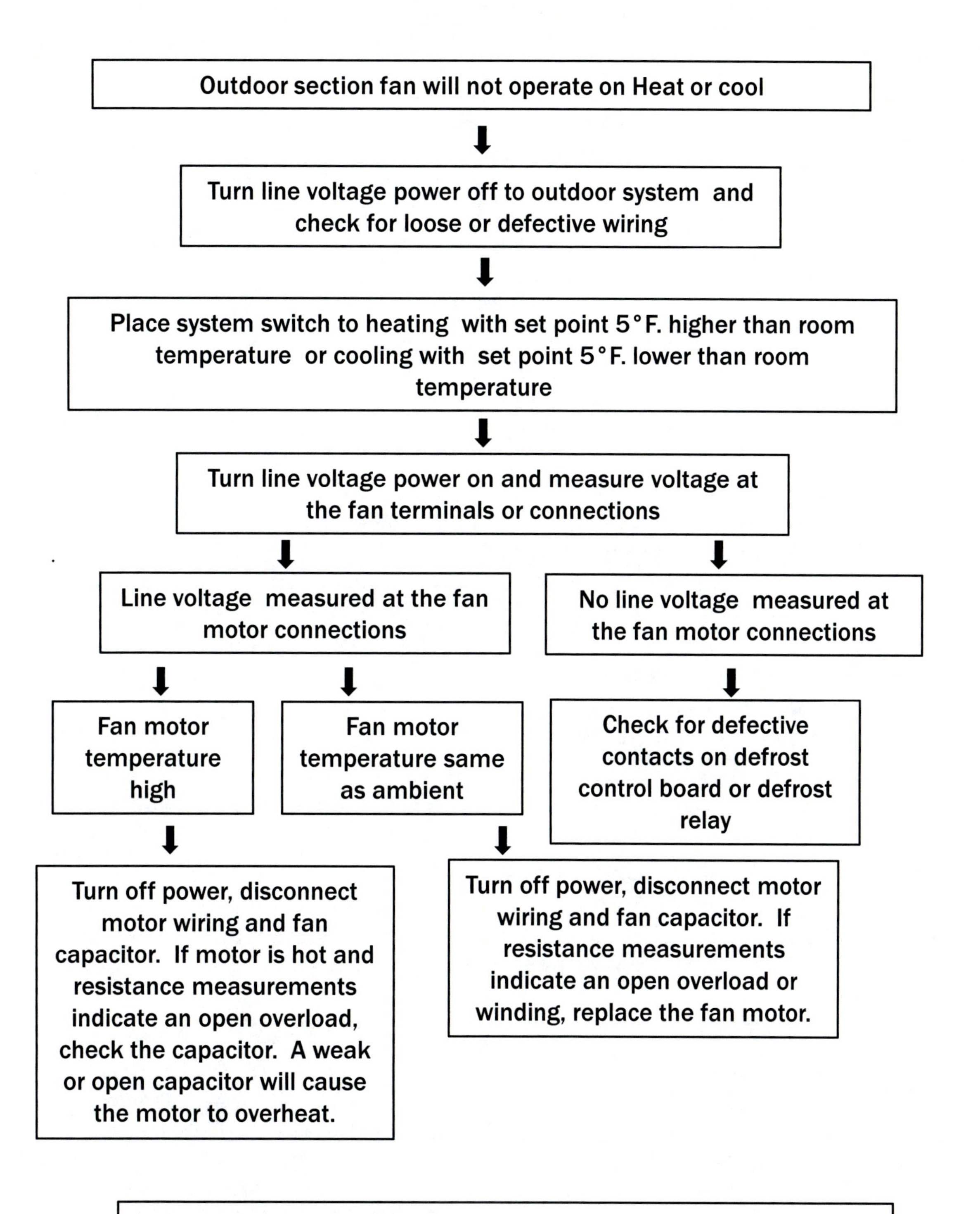

NOTE: Always check the amperage draw of the fan motor. Ampere should not exceed name plate rating.

Heat Pump Troubleshooting Flow Chart #5

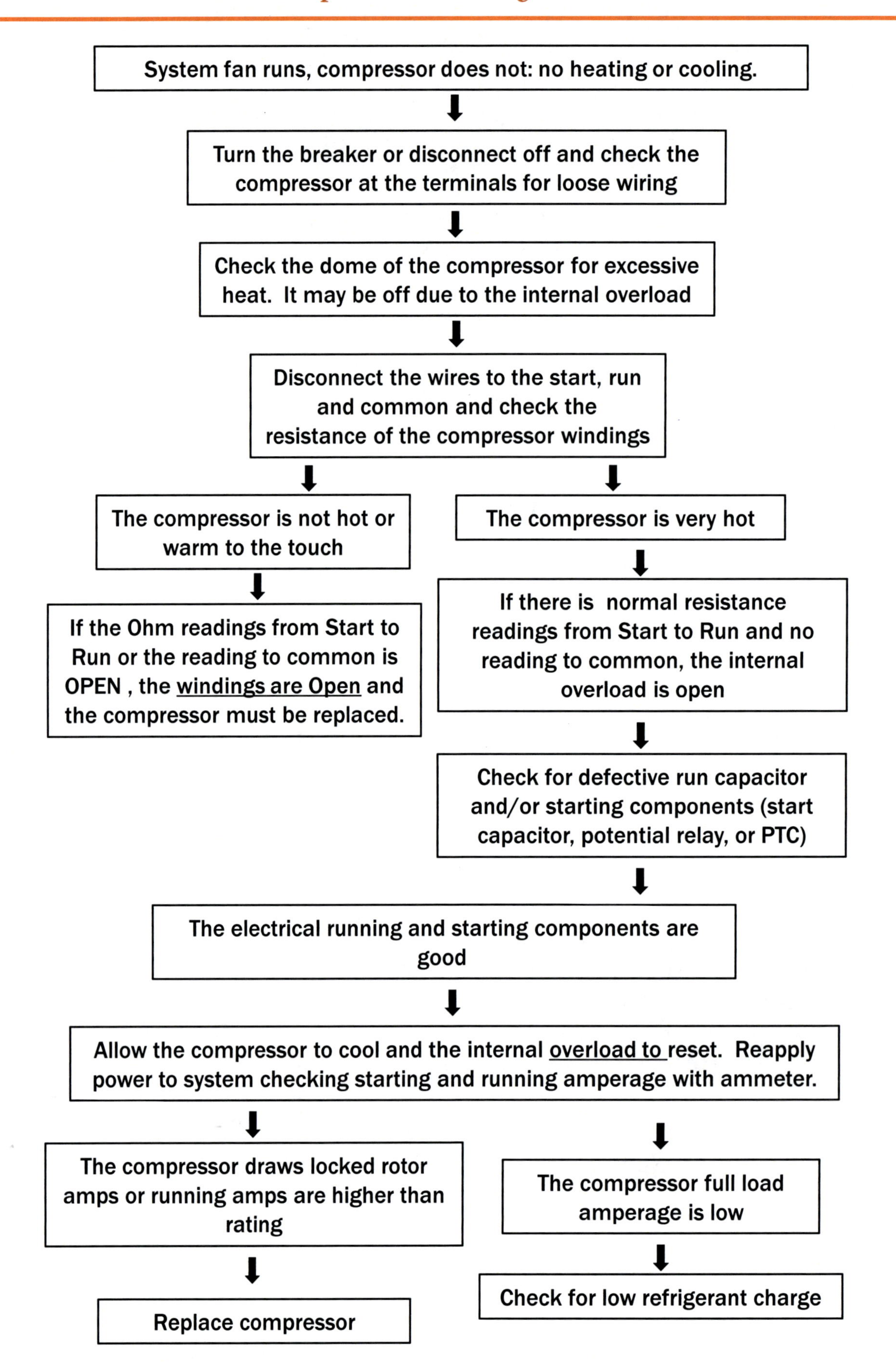

Review Questions
Section 12: Troubleshooting

Name: ______________________________ **Date:** ________________

1. What is a physical sign of a faulty run capacitor?

2. A PTC will have ________ resistance when cold.

3. What should be the maximum temperature difference between the suction lines on a reversing valve?

4. A PSC compressor with an open internal overload with have resistance between the ___________.

5. If the compressor draws high amperes and will not start, always look first for _________________.

6. In a live circuit with the compressor operating, what would a voltage measurement of 10 volts from L1to T1 indicate.

7. Before attempting to service any heat pump, time should always be taken to understand the sequence of operations for the system being addressed. Without a proper understanding of how an individual unit operates, it is nearly impossible to service the unit using a systematic approach. List the five questions to answer that will help pin point problems quickly and easily:

8. What is perhaps the biggest problem with compressors and a major cause of compressor failure?

9. When troubleshooting a heat pump's high voltage controls, the presence of voltage can be determined by placing one lead from the meter to a ________________and moving the other lead from place to place.

10. What would be the appropriate method to check the refrigerant charge?

11. Explain how to read a ladder diagram?

12. What is a pictorial diagram used for?

13. What is a troubleshooting chart best useful for?

14. Explain what is involved in checking the "NET COOLING" for system?

Lab Assignment 1

Section 12: Troubleshooting

Name: ______________________________ **Date:** ________________

Objective: Troubleshoot electric problems for a split system heat pump using Hampden's Heat Pump Trainer model H-HPT-1C.

Directions: Read manufacturers literature and operate trainer without any faults to get familiar with the operation of the system. Using a multimeter, take electrical measurements to troubleshoot the following faults A2, D2, and B5. Read customer complaint, follow manufacturers directions troubleshoot and list repair action required for each fault.

NOTE: Perform the following task using all safety procedures! Input a fault only after the power is turned off to the trainer.

1. Select fault "A2."

   Customer complaint: The system will not come on.

Repair action required: ____

____

1. Select fault "D2."

   Customer complaint: The system will not heat, cold air comes out unless the system is placed in emergency heat.

Repair action required: ____

____

1. Select fault "B5."

   Customer complaint: The system is running but there no cold air coming out.

Repair action required: ____

____

Section 13: Dual-Fuel Systems

Objectives

Upon completion of this section, the participant will be able to:

1. identify characteristics and components of a dual-fuel system
2. understand how dual-fuel systems operate

Dual Fuel Systems

Dual fuel heat pump systems use a heat pump as the primary heating system and a second heating source as auxiliary heat. The most common application uses a fossil fuel furnace as the auxiliary heat source.

Electric strip heat is energized with the heat pump to increase capacity; however, in the case a of a fossil fuel furnace, the heat pump must be turned off before the furnace is energized. The heat pump will run with an extremely high head pressure if both the furnace and heat pump were to operate at the same time.

The indoor coil of the heat pump is installed in the airflow *after* the furnace, due to the cooling cycle. If the coil is placed in the airflow *before* the furnace, the cold moist air leaving the coil during the cooling cycle will corrode the heat exchanger.

Most manufacturers make a control kit that can be installed with a standard heat pump and fossil fuel furnace to create a dual fuel installation. The low voltage control kit is used to switch heat pump operation to furnace operation at the balance point.

The dual fuel control uses mechanical or electronic components to control the operation of both systems. The more advanced controls are electronic and can be used to control multiple stages of heating and cooling.

The balance point temperature setting may depend upon building construction, geographic location, and fuel cost. In most installations, balance point is based on the heat load of the structure. When using **economic balance point** the outdoor thermostat is set to take advantage of fuel cost between the two heating systems. This requires the constant monitoring of fuel cost and is used primarily in commercial applications.

A properly installed dual fuel system allows only the furnace to be turned on at the balance point setting. The second stage of the room thermostat should not be used to switch between heat pump operation and furnace operation without an outdoor thermostat. Using a setback room thermostat without an outdoor thermostat set for the balance point wastes fuel and costs more to operate. Anytime the structure is two or more degrees cooler than the room thermostat set point, the second-stage bulb energizes the auxiliary heat, *unless* there is an outdoor thermostat. Using the second-stage bulb to control the furnace allows the structure to heat faster; however, it is not economical if the furnace is operating on liquid petroleum above the balance point. Most customers using a dual fuel heat pump system with a liquid petroleum fueled furnace combination do so to reduce both the amount of liquid petroleum used and the total energy cost.

The pictorial diagram illustrates a control module with low-voltage wiring for a dual fuel system.

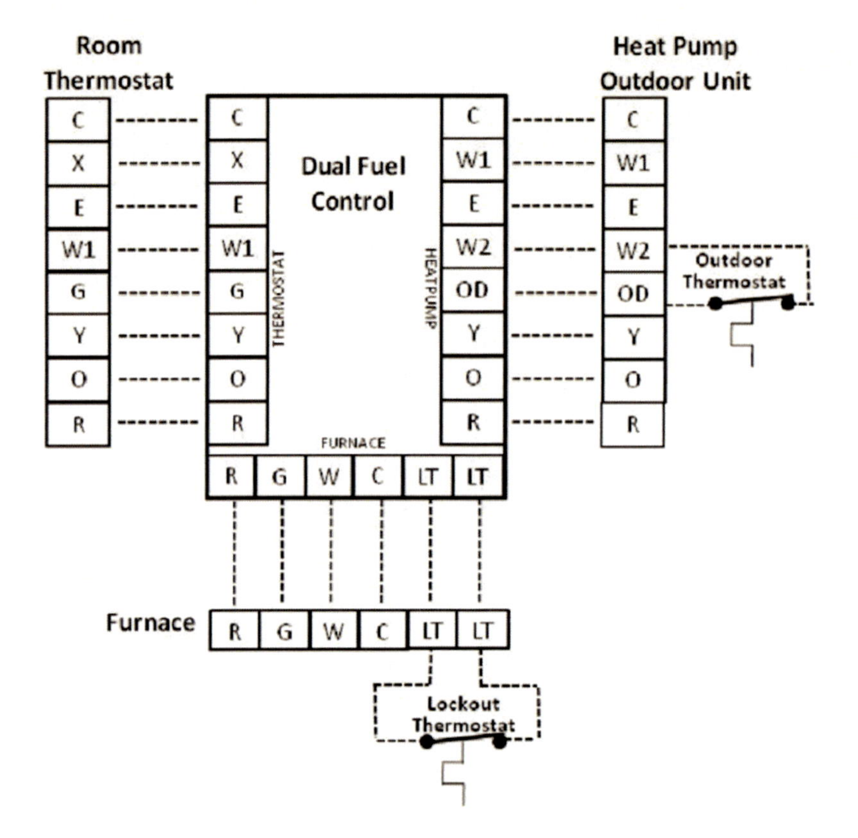

Dual Fuel Add-On Control Kit

Review Questions
Section 13: Dual Fuel

Name: ______________________________ **Date:** ____________________

1. What is the primary heat source for a dual fuel heat pump system?

2. Why is the indoor coil of the heat pump installed in the airflow after the furnace?

3. A properly installed dual fuel system allows only the __________ to be turned on at the balance point setting.

4. Most commonly used dual fuel systems are the heat pump combined with a __________.

5. What is economic balance point?

Section 14: Geothermal Systems

Objectives

Upon completion of this section, the participant will be able to:

1. describe how a geothermal system operates
2. compare direct geothermal to water source heat pump systems
3. identify various piping options

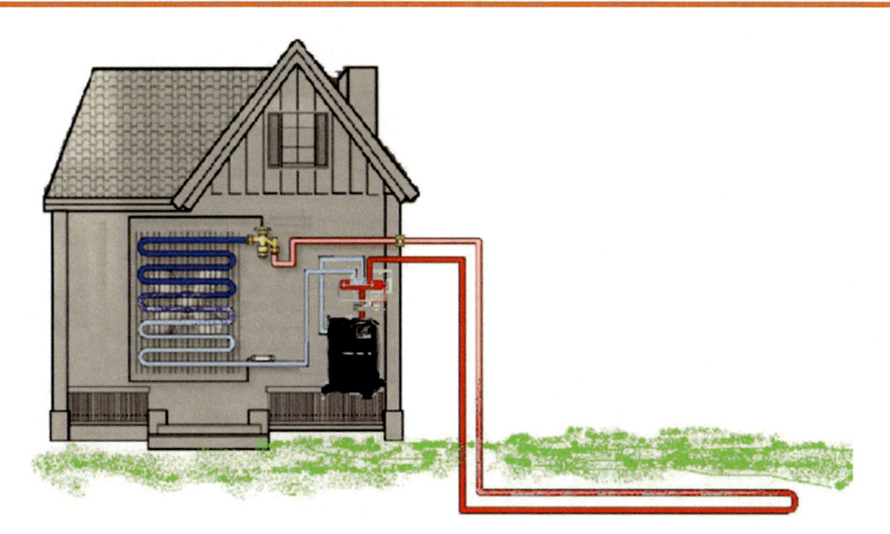

Direct Exchange

Principles of Operation

Geothermal heat pumps have a higher SEER rating than air-source heat pump systems. Using earth as the heat source allows the system to operate at stable temperatures. In northern climates, applications for air-source heat pumps are limited; geothermal or ground-source applications can be used in any climate. As long as the heat exchanger is below the thermal frost line, the ground temperature remains fairly constant.

Water-source heat pumps are package systems with all components housed in one cabinet. The water pump, expansion tank and plastic piping for circulating heat transfer fluid are the only additional components to be installed. The supply and return air duct systems are designed in the same way as any other heating and cooling system.

The control wiring on a water-source heat pump is simpler than on an air-source heat pump. In place of the outdoor coil is a tube-in-a–tube heat exchanger. Unlike in the air-source unit, the coil cannot ice up eliminating the need for a defrost cycle and control, increasing the system efficiency.

The water pump becomes another load which must be accounted for when calculating efficiency; it basically replaces the outdoor fan motor not used on the water-source unit. Many methods are used to create a heat exchange from the heat pump to the earth. One of the simplest is to use water from a lake, pond, river, or stream. Water is pumped from the source to the heat pump's heat exchanger and then to a ditch or back to the source. (When a pond is used, it must have a volume of water equal to or greater than twice the volume of the structure.)

The most common method does not circulate water from the ground but instead uses a glycol solution circulated in a closed loop buried in the ground, in a well hole or pond. The least common method is to bury the refrigerant line in the ground. The first systems to employ this method had problems with oil return to the compressor. Advancements in types of refrigerant lines and heat transfer grouts have reduced oil return problems making the use of this type heat pump more common.

Some models have energy efficiency options such as a heat exchanger that is used to produce domestic hot water. Some systems heat the water when the unit is running in the heating or cooling mode only; others have controls that run the heat pump only as needed to heat the water.

Direct Geothermal

A direct-coupled system configuration circulates refrigerant using copper pipes underground.

The deep well direct heat exchanger has a higher heat transfer rate to the earth compared to other types. This system requires twenty to fifty percent less excavation (drilling, trenching, grouting, etc.) than a properly installed water-source geothermal system. Special consideration must be given to exchanger's design to prevent oil logging which reduces oil return to the compressor.

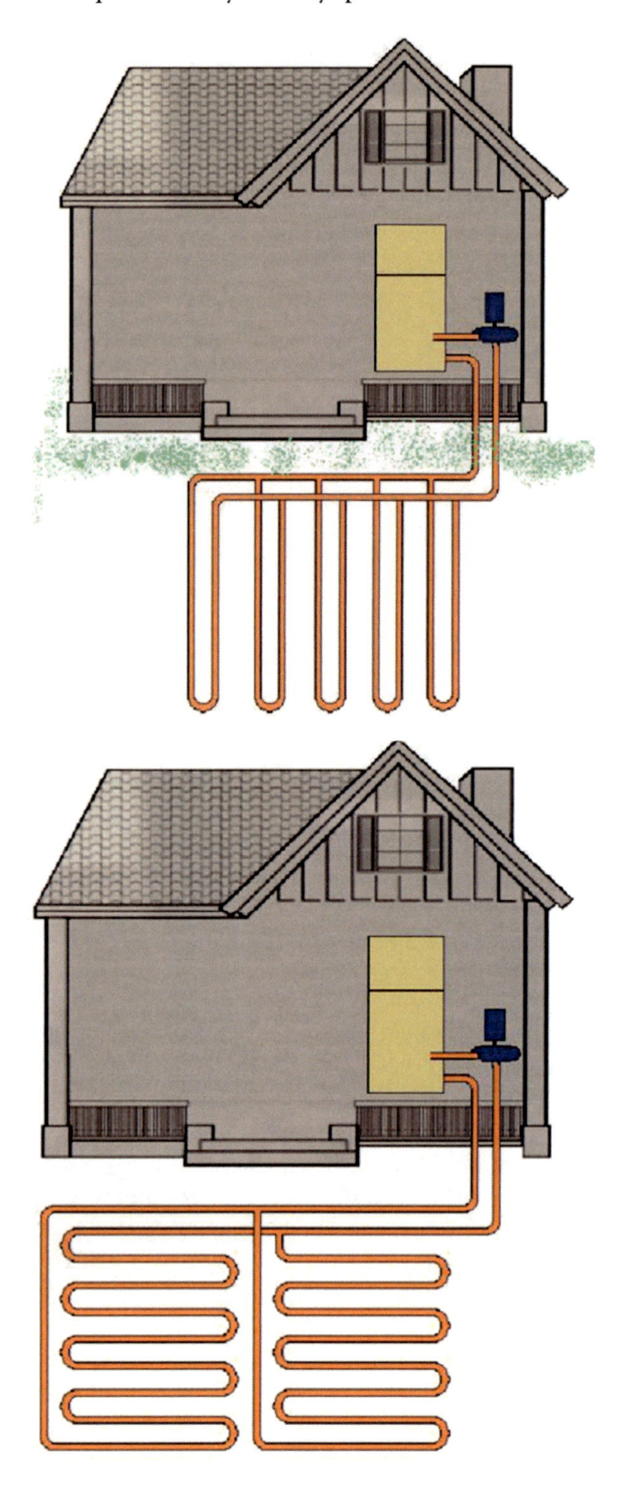

Above: Vertical Loop Below: Horizontal Loop

Water Source

A water-source geothermal system circulates water using polyethylene (PE) pipes buried underground and requires a water circulation pump. The heat transfer fluid pipes, expansion tank and pump are connected to a tube-in-tube heat exchanger. Water-source heat pump systems can be used in numerous applications; the design difference is in the method used to install the pipes in the ground for heat transfer.

Water Geothermal: Vertical Loop

Water circulates inside a loop installed vertically underground. This layout is similar to that of a common well. Water continually circulates in the loop when the heat pump is cooling or heating. A vertical loop requires less physical space than a horizontal loop—approximately 250 feet of vertical loop per 12,000 BTUs in southern United States. Several wells may be connected to meet system capacity needs. Average depth of the wells range from 600 to 800 feet, and wells are spaced approximately ten feet apart.

Although this system is not taking water out of the ground like a water well does, most communities require licenses and permits for drilling bore holes for the plastic pipes.

Water Geothermal: Horizontal Loop

Water circulates inside a loop installed horizontally in the ground. The length of the loop is determined by soil type, depth, and system BTU capacity— approximately 400 feet per 12,000 BTUs in the southern United States. Water circulates continuously in a closed loop to move heat between the heat pump and the ground. This system requires enough space to install the designed length of horizontal loop. In most

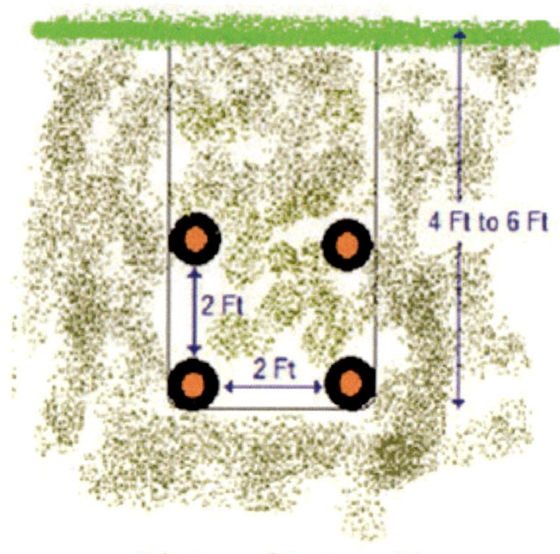

Side View of Horizontal Loop

applications, four lines or pipes are buried in one two-foot-wide by six-foot-deep trench.

Open-Loop Systems

Open-loop systems use one or more wells or surface body water as the heat exchange fluid that circulates directly through the heat exchanger. After the water has circulated through the system, it returns to the ground through the well, recharge well, or surface discharge. Open-loop systems can only be used in locations with an adequate supply of clean water. Follow all local codes and federal regulations regarding groundwater discharge.

Piping

Plastics such as polyethylene or polybutylene are common piping materials for geothermal heat ground-heat exchangers. The method for connecting polyethylene pipes underground is butt fusion; for polybutylene pipes it is socket fusion.

The length of the heat exchanger needed varies with geographic location, due to varying soil conditions. The higher the ground water table, the better the heat transfer. For horizontal loop installations, there may be a great difference in lengths of piping required. In some horizontal runs, the piping is coiled like a spring in a trench.

Plastic tubing may be coiled or wrapped in bundles with weights attached and sunk to the bottom of a pond, lake or river.

Review Questions
Section 14: Geothermal Systems

Name: ______________________________ **Date:** __________________

1. Geothermal heat pumps have a ________ SEER rating than air-source heat pump systems.

2. A ___________ geothermal system circulates water using polyethylene (PE) pipes buried underground and requires a water circulation pump.

3. A _______________ geothermal system configuration circulates refrigerant using copper pipes underground.

4. _______________systems use one or more wells or surface body water as the heat exchange fluid that circulates directly through the heat exchanger.

5. What type of piping is used in conjunction with water source heat pumps?

6. In a vertical loop system, approximately how many feet of well should be used per ton of capacity?

7. A horizontal loop system will normally have __________ pipes buried in a 2 foot x 6 foot deep trench.

8. Does a geothermal heat pump need a defrost cycle?